Arduino Wearable Projects

Design, code, and build exciting wearable projects using Arduino tools

Tony Olsson

PUBLISHING

BIRMINGHAM - MUMBAI

Arduino Wearable Projects

First published: August 2015

Production reference: 1250815

Published by Packt Publishing Ltd.
Livery Place
35 Livery Street
Birmingham B3 2PB, UK.

ISBN 978-1-78528-330-7

www.packtpub.com

Credits

Author
Tony Olsson

Reviewers
Tomi Dufva

Kristina Durivage

Jimmy Hedman

Kallirroi Pouliadou

Gabriela T. Richard

Johnty Wang

Commissioning Editor
Priya Singh

Acquisition Editor
Vivek Anantharaman

Content Development Editor
Pooja Nair

Project Coordinator
Suzanne Coutinho

Technical Editor
Rupali R. Shrawane

Copy Editor
Charlotte Carneiro

Proofreader
Safis Editing

Indexer
Rekha Nair

Production Coordinator
Manu Joseph

Cover Work
Manu Joseph

About the Author

Tony Olsson works as a lecturer at the University of Malmö, where he teaches multiple design fields with the core being physical prototyping and wearable computing. His research includes haptic interactions and telehaptic communication. Olsson has a background in philosophy and traditional arts, but later shifted his focus to interaction design and computer science. He is also involved in running the IOIO laboratory at Malmö University.

Besides his work at the university, he also works as a freelance artist/designer and author. Prior to this publication, Olsson published two books based on wearable computing and prototyping with Arduino and Arduino-based platforms.

I would like to thank all the people and students of the IOIO laboratory and the K3 institution, both current and past. The work we do together has always been inspiring. Thanks to my sister and mother for all their support. A special thanks to David Cuartielles and Andreas Göransson. Without our endeavors together, this book probably would have never been written. I would also like to thank Hemal and Pooja at Packt; it has been a true pleasure working with them on this book. I'd also like to thank the rest of the Arduino team, Massimo Banzi, David Mellis, and Tom Igoe, for their impressive work with Arduino; and the Arduino community, which remains the best in the world. Last but not least, I would like to thank Jennie, I can only hope to repay all the support and understanding she has given me during the process of writing this book.

About the Reviewers

Tomi Dufva is an MA in fine arts and a doctoral researcher at Aalto ARTS University. He is a cofounder of Art and Craft School Robotti and lives and works in Turku as a visual artist, art teacher, and researcher. Tomi researches creative coding at Aalto University, in the school of Arts, Design, and Architecture. Tomi specializes in code literacy, maker culture, pedagogical use of code, and integrating painting and drawing with electronics and code. Tomi has taught in schools from kindergartens to universities. You can see Tomi's research on his blog (www.thispagehassomeissues.com).

Kristina Durivage is a software developer by day and hardware hacker by night. She is well-known for her TweetSkirt—an item of clothing that displays tweets. She lives in Minneapolis, Minnesota, and can be found on Twitter at @gelicia.

Jimmy Hedman is a professional HPC (High Performance Computing) geek who works with large systems where size is measured by the number of racks and thousands of cores. In his spare time, he goes in the opposite direction and focuses on smaller things, such as Beaglebone Blacks and Arduinos.

He is currently employed by South Pole AB, the biggest server manufacturer in Sweden, where he is a Linux consultant with HPC as his main focus.

He has previously reviewed *Arduino Robotics Projects* for Packt Publishing.

> I would like to thank my understanding wife, who lets me go on with my hobbies like I do. I also would like to thank Packt Publishing for letting me have this much fun with interesting stuff to read and review.

Kallirroi Pouliadou is an interaction designer with a strong visual design and architecture background, and experience in industrial design, animation, and storytelling. She explores technology as an amateur maker.

Johnty Wang has a masters of applied science degree in electrical and computer engineering from the University of British Columbia. His main area of research is developing New Interfaces for Musical Expression (NIME), and it is supported by his personal passion for music and human-technology interfaces. He has a diverse range of experience in hardware and software systems, developing embedded, mobile, and desktop applications for works ranging from interactive installations to live musical performances. His work has appeared at festivals, conferences, and competitions internationally. Johnty is currently a PhD student in music technology at McGill University, supervised by professor Marcelo Wanderley.

www.PacktPub.com

Support files, eBooks, discount offers, and more

For support files and downloads related to your book, please visit www.PacktPub.com.

Did you know that Packt offers eBook versions of every book published, with PDF and ePub files available? You can upgrade to the eBook version at www.PacktPub.com and as a print book customer, you are entitled to a discount on the eBook copy. Get in touch with us at service@packtpub.com for more details.

At www.PacktPub.com, you can also read a collection of free technical articles, sign up for a range of free newsletters and receive exclusive discounts and offers on Packt books and eBooks.

https://www2.packtpub.com/books/subscription/packtlib

Do you need instant solutions to your IT questions? PacktLib is Packt"s online digital book library. Here, you can search, access, and read Packt"s entire library of books.

Why subscribe?

- Fully searchable across every book published by Packt
- Copy and paste, print, and bookmark content
- On demand and accessible via a web browser

Free access for Packt account holders

If you have an account with Packt at www.PacktPub.com, you can use this to access PacktLib today and view 9 entirely free books. Simply use your login credentials for immediate access.

Table of Contents

Preface

Almost 10 years have passed since I picked up my first Arduino board. At the time, I was an interaction design student at Malmö University. At the front of the classroom that day, there was a bearded Spaniard talking, rather claiming, that he could teach us all about electronics and how to do programming for microprocessors, all in 1 week. Of course, since I knew nothing about electronics and never thought I would learn anything about it, I did not believe him.

The Spaniard had a completely new approach to teaching, which I had never encountered before. He wanted to teach us, not by books and lectures, but by doing things. One of my classmates pointed out that most of us did not know anything about electronics, so how are we supposed to do anything with it? The Spaniard replied that it does not matter, you can do things without knowing what you are doing, and by doing them, you will learn.

After 15 minutes, we all had connected a small lamp to our Arduino boards, and we had managed to program the lamp so that it would turn itself on and off. What baffled me was not only what we had achieved in such little time, but also that parts of what was going on actually made sense. We were learning by doing.

The bearded Spaniard was actually David Cuartielles, who together with Massimo Banzi, just 1 year before, invented the Arduino board. Soon after they invented it, Tome Igoe and David Mellis joined the team, and as they say, the rest is history. But I still remember that day, as if it was yesterday, when I looked down at my blinking light and something sparked inside me. I wanted to learn and do more. Then David gave me the second valuable lesson, that the best way to learn more is to share your knowledge with others, and he put me in a position where I was able to do so. Again I was skeptical, since I had no knowledge to speak of, but again the lesson followed, even if you only know a little, it is enough to help those that know nothing yet.

Soon after, I found out about a field called wearable computing. The idea was to design and apply a technology to the human body in different ways, and it all sounded as wonderfully crazy as the idea that you could learn electronics and programming without any prior knowledge of how to do so. With inspiration from Arduino and its team members, I leaped headfirst into the field. In this new field, I found new inspiration in the works of Steve Mann and Leah Buechley. Mann, now a professor at the University of Toronto, developed his own wearable computer in the 80s and had mostly done so on his own. Buechley, also a professor at MIT, had taken the Arduino board and developed a new prototyping platform, which is specialized for a wearable context. Both seemed to have done this against all the odds. Again, I was inspired, and started to develop my own wearable devices, teaching others how to do the same. Eventually, I collected enough know-how on things that I started to write them down. When I started to share my writing, I found out how truly amazing the Arduino community is a world-wide group of people that share a love for making things with electronics.

It's safe to say that if it had not been for all these people, I probably would never have written any of my books, so I would like to extend my thanks to all. I would also like to thank you for picking up this book. You might be a novice or an expert, but I do hope it will not matter. This book is based on the idea that anyone can learn anything by the simple principle of actually "doing." If you are already an expert, then you know there is always something to learn from "doing" things in a new way.

So, I hope you will gain some new knowledge and inspiration from the projects we created in this book, and I wish you all the best in your creating endeavors.

Do check out "Soldering with David Cuartielles" on my YouTube channel at `https://www.youtube.com/watch?v=Mg01HFjsn6k`.

What this book covers

Chapter 1, *First Look and Blinking Lights*, covers the basic steps of installing the development environment and how to get started with coding. We also take a look at how to create our first circuit and control an LED.

Chapter 2, *Working with Sensors*, teaches about interfacing with sensors and extracting data from them. The chapter also introduces digital and analog sensors ranging from simple to complex sensors.

Chapter 3, *Bike Gloves*, introduces the reader to the first project of the book, where the goal is to create a pair of bike gloves. In this chapter, we introduce the use of LEDs and how to control them, as well as how to use sensors for some simple gesture recognition.

Chapter 4, LED Glasses, teaches you to create a pair of programmable LED glasses. These glasses will be covered by LEDs in the front, which will be programmable to display different patterns and shapes. The reader will also be introduced to the construction of a pair of sunglasses.

Chapter 5, Where in the World Am I?, focuses on the making of a wrist-worn GPS tracking device. The information will be displayed on a small LCD screen. This chapter also includes instructions and tips on how to create a casing containing the components so that the device can be worn on the wrist.

Chapter 6, Hands-on with NFC, deals with NFC technology and servomotors and how they can be combined into a smart door lock. This chapter also includes how to design around NFC tags and make wearable jewelry that will work as a key for the lock.

Chapter 7, Hands-on BLE, deals with low-powered Bluetooth technology and how it can be implemented into wearable projects. This chapter introduces the Blend Micro board and how it can be used to create projects that connect to your mobile phone.

Chapter 8, On the Wi-fly, introduces you to the Wi-Fi Particle Core board and its web IDE. This chapter also talks about how to connect to online services.

Chapter 9, Time to Get Smart, focuses on the creation of a smart watch, which connects to the Internet and uses online services to create custom notifications to be displayed on a small OLED screen.

The online chapter (*Chapter 10*), *Interactive Name Tag,* expands upon *Chapter 7, Hands-on BLE,* which deals with small screens, and shows you how to interact with them over Bluetooth in order to make an interactive name tag. This chapter is available at `https://www.packtpub.com/sites/default/files/downloads/ArduinoWearableProjects_OnlineChapter.pdf`.

What you need for this book

Download and install the preconfigured Arduino IDE from Adafruit: `https://learn.adafruit.com/getting-started-with-flora/download-software`.

The Particle Build Web IDE, sign up for a free account on: `https://build.particle.io/login`.

Free account on IFTTT: `https://ifttt.com/`.

Boards

Here's a list of the boards you'll work on:

- Adafruit Trinket—Mini Microcontroller—5V Logic
- Adafruit Pro Trinket—5V 16 MHz
- FLORA—Wearable electronic platform: Arduino-compatible
- Spark Core with Chip Antenna Rev 1.0
- Redbear Blend Micro BLE board

Components and tools

Here's a list of all the components and tools you need:

- Soldering iron
- GA1A12S202 Log-scale Analog Light Sensor
- Long Flex/Bend sensor
- LDRs
- Adafruit TSL2561 Digital Luminosity/Lux/Light Sensor Breakout
- Breadboarding wire bundle
- Flora Wearable Ultimate GPS Module
- Monochrome 128 x 32 I2C OLED graphic display
- Adafruit LED Sequins
- 3.56 MHz RFID/NFC tags
- Adafruit PN532 NFC/RFID Controller Shield for Arduino + Extras
- Lithium Ion Polymer Battery—3.7V 1200 mAh
- SHARP Memory Display Breakout—1.3" 96 x 96 Silver Monochrome
- Small Alligator Clip Test Lead
- Lithium Ion Polymer Battery—3.7V 500mAh
- Monochrome 1.3" 128x64 OLED graphic display
- Adafruit Micro Lipo w/MicroUSB Jack—USB LiIon/LiPoly charger (V1)
- Full-sized breadboard
- OLED Breakout Board—16-bit Color 0.96" w/microSD holder
- Half-sized breadboard
- USB cable—6" A/MiniB

- FLORA 9-DOF Accelerometer/Gyroscope/Magnetometer – LSM9DS0 (V1.0)
- Lithium Ion Polymer Battery – 3.7V 150mAh
- Hook-up Wire Spool Set – 22AWG Solid Core – 6 x 25 ft
- Flush diagonal cutters
- Helping Third Hand Magnifier W/Magnifying Glass Tool

Who this book is for

For readers familiar with the Arduino prototyping platform with some prior experienced with ordinary hardware tools.

Conventions

In this book, you will find a number of text styles that distinguish between different kinds of information. Here are some examples of these styles and an explanation of their meaning.

Code words in text, database table names, folder names, filenames, file extensions, pathnames, dummy URLs, user input, and Twitter handles are shown as follows: "We can include other contexts through the use of the include directive."

A block of code is set as follows:

```
//Variable to store the pin
int ldrSensor = 10;

void setup(){
//Start the serial communication
  Serial.begin(9600);
}

void loop(){
//Save the data from the sensor into storeData
  int storeData=analogRead(ldrSensor);
//Re-map storeData to a new range of values
  int mapValue=map(storeData,130,430,0,2000);
//Print the re-mapped value
  Serial.println(mapValue);
//Give the computer some time to print
  delay(200)
}
```

New terms and **important words** are shown in bold. Words that you see on the screen, for example, in menus or dialog boxes, appear in the text like this: "Clicking the **Next** button moves you to the next screen."

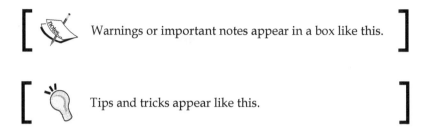

Warnings or important notes appear in a box like this.

Tips and tricks appear like this.

Reader feedback

Feedback from our readers is always welcome. Let us know what you think about this book—what you liked or disliked. Reader feedback is important for us as it helps us develop titles that you will really get the most out of.

To send us general feedback, simply e-mail feedback@packtpub.com, and mention the book's title in the subject of your message.

If there is a topic that you have expertise in and you are interested in either writing or contributing to a book, see our author guide at www.packtpub.com/authors.

Customer support

Now that you are the proud owner of a Packt book, we have a number of things to help you to get the most from your purchase.

Downloading the example code

You can download the example code files from your account at http://www.packtpub.com for all the Packt Publishing books you have purchased. If you purchased this book elsewhere, you can visit http://www.packtpub.com/support and register to have the files e-mailed directly to you.

Downloading the color images of this book

We also provide you with a PDF file that has color images of the screenshots/diagrams used in this book. The color images will help you better understand the changes in the output. You can download this file from `https://www.packtpub.com/sites/default/files/downloads/ArduinoWearableProjects_ColorImages.pdf`.

Errata

Although we have taken every care to ensure the accuracy of our content, mistakes do happen. If you find a mistake in one of our books — maybe a mistake in the text or the code — we would be grateful if you could report this to us. By doing so, you can save other readers from frustration and help us improve subsequent versions of this book. If you find any errata, please report them by visiting `http://www.packtpub.com/submit-errata`, selecting your book, clicking on the **Errata Submission Form** link, and entering the details of your errata. Once your errata are verified, your submission will be accepted and the errata will be uploaded to our website or added to any list of existing errata under the Errata section of that title.

To view the previously submitted errata, go to `https://www.packtpub.com/books/content/support` and enter the name of the book in the search field. The required information will appear under the **Errata** section.

Piracy

Piracy of copyrighted material on the Internet is an ongoing problem across all media. At Packt, we take the protection of our copyright and licenses very seriously. If you come across any illegal copies of our works in any form on the Internet, please provide us with the location address or website name immediately so that we can pursue a remedy.

Please contact us at `copyright@packtpub.com` with a link to the suspected pirated material.

We appreciate your help in protecting our authors and our ability to bring you valuable content.

Questions

If you have a problem with any aspect of this book, you can contact us at `questions@packtpub.com`, and we will do our best to address the problem.

1
First Look and Blinking Lights

The basis for this book is the Arduino platform, which refers to three different things: software, hardware, and the Arduino philosophy. The hardware is the Arduino board, and there are multiple versions available for different needs. In this book, we will be focusing on Arduino boards that were made with wearables in mind. The software used to program the boards is also known as the **Arduino IDE**. IDE stands for **Integrated Development Environment**, which are programs used to write programs in programming code. The programs written for the board are known as **sketches**, because the idea aids how to write programs and works similar to a sketchpad. If you have an IDE, you can quickly try it out in code. This is also a part of the Arduino philosophy. Arduino is based on the open source philosophy, which also reflects on how we learn about Arduino. Arduino has a large community, and there are tons of projects to learn from.

First, we have the Arduino hardware, which we will use to build all the examples in this book along with different additional electronic components. When the Arduino projects started back in 2005, there was only one piece of headwear to speak of, which was the serial Arduino board. Since then, there have been several iterations of this board, and it has inspired new designs of the Arduino hardware to fit different needs. If you are familiar with Arduino for a while, you probably started out with the standard Arduino board. Today, there are different Arduino boards that fit different needs, and there are countless clones available for specific purposes. In this book, we will be using different specialized Arduino boards such as the FLORA board and Spark core board.

The Arduino software that is Arduino IDE is what we will use to program our projects. The IDE is the software used to write programs for the hardware. Once a program is compiled in the IDE, it will upload it to the Arduino board, and the processor on the board will do whatever your program says. Arduino programs are also known as sketches. The name sketches is borrowed from another open source project and software called **Processing**. Processing was developed as a tool for digital artists, where the idea was to use Processing as a digital sketchpad.

The idea behind sketches and other aspects of Arduino is what we call the Arduino philosophy, and this is the third thing that makes Arduino. Arduino is based on open source, which is a type of licensing model where you are free to develop you own designs based on the original Arduino board. This is one of the reasons why you can find so many different models and clones of the Arduino boards. Open source is also a philosophy that allows ideas and knowledge to be shared freely. The Arduino community has grown strong, and there are many great resources to be found, and Arduino friends to be made.

The only problem may be where to start? Books like this one are good for getting you started or developing skills further. Each chapter in this book is based on a project that will take you from the start, all the way to a finished "prototype". I call all the project prototypes because these are not finished products. The goal of this book is also for you to develop these projects further, once you have completed the chapter. As your knowledge progresses, you can develop new sketches to run on you prototypes, develop new functions, or change the physical appearance to fit your needs and preferences.

In this chapter, you will have a look at:

- Installing the IDE
- Working with the IDE and writing sketches
- The FLORA board layout
- Connecting the FLORA board to the computer
- Controlling and connecting LEDs to the FLORA board

Wearables

This book is all about wearables, which are defined as computational devices that are worn on the body. A computational device is something that can make calculations of any sort. Some consider mechanical clocks to be the first computers, since they make calculations on time. According to this definition, wearables have been around for centuries, if you think about it. Pocket watches were invented in the 16th century, and a watch is basically as small device that calculates time. Glasses are also an example of wearable technology that can be worn on your head, which have also been around for a long time. Even if glasses do not fit our more specified definition of wearables, they serve as a good example of how humans have modified materials and adapted their bodies to gain new functionality. If we are cold, we dress in clothing to keep us warm, if we break a leg, we use crutches to get around, or even if an organ fails, we can implant a device that replicates their functionality. Humans have a long tradition of developing technology to extend the functionality of the human body.

With the development of technology for the army, health care, and professional sport, wearables have a long tradition. But in recent years, more and more devices have been developed for the consumer market. Today, we have smart watches, smart glasses, and different types of smart clothing.

In this book, we will carry on this ancient tradition and develop some wearable projects for you to learn about electronics and programming. Some of these projects are just for fun and some have a specific application. The knowledge presented in all the chapters of this book progresses from the chapter before it. We will start off slow, and the chapters will gradually become more complex both in terms of hardware and software. If you are already familiar with Arduino, you can pick any project and get started. If you find it too hard, you can always go back and take a look at the chapter that precedes it. If you're completely new to Arduino, continue reading this chapter as we will go through the installation process of the Arduino IDE and how to get started with programming.

Installing and using software

The projects in this book will be based on different boards made by the company Adafruit. Later in this chapter, we will take a look at one of these boards, called the **FLORA**, and explain the different parts. These boards come with a modified version of the Arduino IDE, which we will be using in the chapter. The Adafruit IDE looks exactly the same as the Arduino IDE. The FLORA board, for example, is based on the same microprocessor as the Arduino Leonardo board and can be used with the standard Arduino IDE but programmed using the Leonardo board option. With the use of the Adafruit IDE the FLORA board is properly named. In this book, we will use two other models called the Gemma and Trinket boards, which are based on a microprocessor that is different from the standard Arduino boards. The Adafruit version of the IDE comes preloaded with the necessary libraries for programming these boards, so there is no need to install them separately.

For downloading and instructions on installing the IDE, head over to the Adafruit website and follow the steps on the website:

```
https://learn.adafruit.com/getting-started-with-flora/download-
software
```

Make sure to download the software corresponding to your operating system. The process for installing the software depends on your operating system. These instructions may change over time and may be different for different versions of the operating system. The installation is a very straightforward process if you are working with OS X. On Windows, you will need to install some additional USB drivers. The process for installing on Linux depends on which distribution you are using. For the latest instructions, take a look at the Arduino website for the different operating systems.

The Arduino IDE

On the following website, you can find the original Arduino IDE if you need it in the future. In this book, you will be fine sticking with the Adafruit version of the IDE, since the most common original Arduino boards are also supported. The following is the link for downloading the Arduino software: `https://www.arduino.cc/en/Main/Software`.

First look at the IDE

The IDE is where we will be doing all of our programming. The first time you open up the IDE, it should look like *Figure 1.1*:

Figure 1.1: The Arduino IDE

The main white area of the IDE is blank when you open a new sketch, and this is the area of the IDE where we will write our code later on. First, we need to get familiar with the functionality of the IDE.

At the top left of the IDE, you will find five buttons. The first one, which looks like a check sign, is the compile button. When you press this button, the IDE will try to compile the code in your sketch, and if it succeeds, you will get a message in the black window at the bottom of you IDE that should look similar to this:

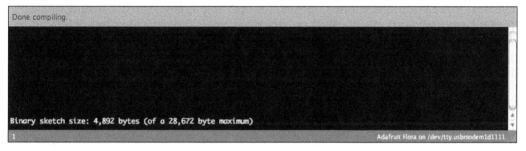

Figure 1.2: The compile message window

When writing code in an IDE, we will be using what is known as a third-level programming language. The problem with microprocessors on Arduino boards is that they are very hard to communicate with using their native language, and this is why third-level languages have been developed with human readable commands. The code you will see later needs to be translated into code that the Arduino board understands, and this is what is done when we compile the code. The compile button also makes a logical check of your code so that it does not contain any errors. If you have any errors, the text in the black box in the IDE will appear in red, indicating the line of code that is wrong by highlighting it in yellow. Don't worry about errors. They are usually misspelling errors and they happen a lot even to the most experienced programmers. One of the error messages can be seen in the following screenshot:

```
expected ';' before 'digitalWrite'

Blink.ino: In function 'void loop()':
Blink:22: error: expected ';' before 'digitalWrite'
```

Figure 1.3: Error message in the compile window

Adjacent to the compile button, you will find the **Upload** button. Once this button is pressed, it does the same thing as the compile button, and if your sketch is free from errors, it will send the code from your computer to the board:

Figure 1.4: The quick buttons

The next three buttons are quick buttons for opening a new sketch, opening an old sketch, or saving your sketch. Make sure to save your sketches once in a while when working on them. If something happens and the IDE closes unexpectedly, it does not autosave, so manually saving once in a while is always a good idea.

At the far right of the IDE you will find a button that looks like a magnifying glass. This is used to open the **Serial monitor**. This button will open up a new window that lets you see the communication form from, and to, the computer and the board. This can be useful for many things, which we will have a closer look at in *Chapter 2, Working with Sensors*.

At the top of the screen you will find a classic application menu, which may look a bit different depending on your operating system, but will follow the same structure. Under **File**, you will find the menu for opening your previous sketches and different example sketches that come with the IDE, as shown in *Figure 1.5*. Under **Edit**, you will find different options and quick commands for editing your code. In **Sketch**, you can find the same functions as in the buttons in the IDE window:

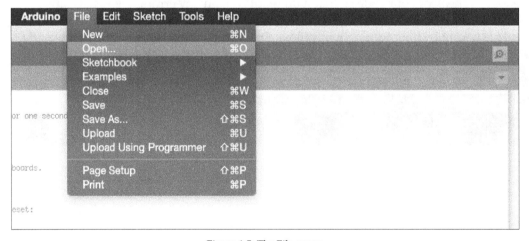

Figure 1.5: The File menu

Under **Tools**, you will find two menus that are very important to keep track of when uploading sketches to your board. Navigate to **Tools | Board** and you will find many different types of Arduino boards. In this menu, you will need to select the type of board you are working with. Under **Tools | Serial port**, you will need to select the USB port which you have connected to your board. Depending on your operating system, the port will be named differently. In Windows, they are named **COM***. On OS X, they are named **/dev/tty.****:**

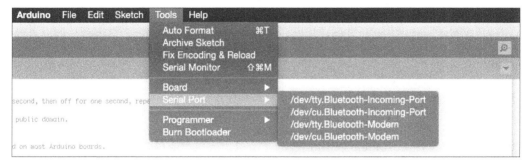

Figure 1.6: The Tools menu

Since there may be other things inside your computer also connected to a port, these will also show up in the list. The easiest way to figure out which port is connected to your board is to:

1. Plug you board in to your computer using a USB cable.
2. Then check the Serial port list and remember which port is occupied.
3. Unplug the board and check the list again.
4. The board missing in the list is the port where your board is connected. Plug your board back in and select it in the list. All Arduino boards connected to you computer will be given a new number.

In most cases, when your sketch will not upload to you board, you have either selected the wrong board type or serial port in the tools menu.

Getting to know you board

As I mentioned earlier, we will not be using the standard Uno Arduino boards in this book, which is the board most people think of when they hear Arduino board. Most Arduino variations and clones use the same microprocessors as the standard Arduino boards, and it is the microprocessors that are the heart of the board. As long as they use the same microprocessors, they can be programmed as normal by selecting the corresponding standard Arduino board in the **Tools** menu. In our case, we will be using a modified version of the Arduino IDE, which features the types of boards we will be using in this book. What sets other boards apart from the standard Uno Arduino boards is usually the form factor of the board and pin layout. In this book, we will be using a board called the FLORA. This board was created with wearables in mind. The FLORA is based on the same chip used in the Arduino Leonardo board, but uses a much smaller form factor and has been made round to ease the use in a wearable context. You can complete all the projects using most Arduino boards and clones, but remember that the code and construction of the project may need some modifying.

The FLORA board

In the following *Figure 1.7* you will find the FLORA board:

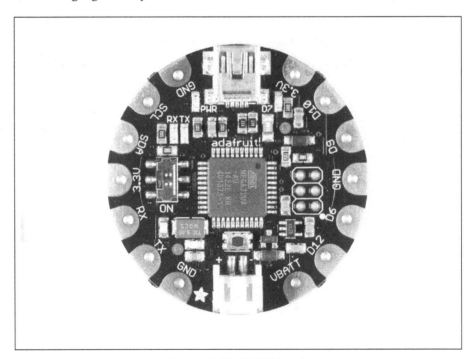

Figure 1.7: The FLORA board

The biggest difference to normal Arduino boards besides the form factor is the number of pins available. The pins are the copper-coated areas at the edge of the FLORA. The form factor of the pins on FLORA boards is also a bit different from other Arduino boards. In this case, the pin holes and soldering pads are made bigger on FLORA boards so they can be easily sewn into garments, which is common when making wearable projects. The larger pins also make it easier to prototype with alligator clips, which we will be using later on in this chapter as shown in *Figure 1.10*. The pins available on the FLORA are as follows, starting from the right of the USB connector, which is located at the top of the board in the preceding *Figure 1.7*:

The pins available on the FLORA are as follows, starting from the right of the USB connector, which is located at the top of the board in *Figure 1.7*:

- **3.3V**: Regulated 3.3 volt output at a 100mA max
- **D10**: Is both a digital pin **10** and an analog pin **10** with PWM
- **D9**: Is both a digital pin **9** and an analog pin **9** with PWM
- **GND**: Ground pin
- **D6**: Is both a digital pin **6** and an analog pin **7** with PWM
- **D12**: Is both a digital pin **12** and an analog pin **11**
- **VBATT**: Raw battery voltage, can be used for as battery power output
- **GND**: Ground pin
- **TX**: Transmission communication pin or digital pin **1**
- **RX**: Receive communication pin or digital pin **0**
- **3.3V**: Regulated 3.3 volt output at a 100mA max
- **SDA**: Communication pin or digital pin **2**
- **SCL**: Clock pin or digital pin **3** with PWM

As you can see, most of the pins have more than one function. The most interesting pins are the **D*** pins. These are the pins we will use to connect to other components. These pins can either be a digital pin or an analog pin. Digital pins operate only in **1** or **0**, which mean that they can be only **On** or **Off**. You can receive information on these pins, but again, this is only in terms of on or off.

The pins marked **PWM** have a special function, which is called **Pulse Width Modulation**. On these pins, we can control the output voltage level. The analog pins, however, can handle information in the range from **0** to **1023**. As we introduce analog sensors in *Chapter 2*, *Working with Sensors*, we will look into the differences between them in more detail.

The **3.3V** pins are used to power any components connected to the board. In this case, an electronic circuit needs to be completed, and that's why there are two **GND** pins. In order to make an electronic circuit, power always needs to go back to where it came from. For example, if you want to power a motor, you need power from a power source connected via a cable, with another cable directing the power back to the power source, or the motor will not spin.

TX, **RX**, **SDA**, and **SCL** are pins used for communication, which we will have a look at later on in the book in the chapters dealing with more complex sensors. The **VBATT** pin can be used to output the same voltage as your power source, which you connect to the connector located at the bottom of the FLORA board shown in *Figure 1.7*.

Other boards

In *Figure 1.8* you will find the other board types we will be using in this book:

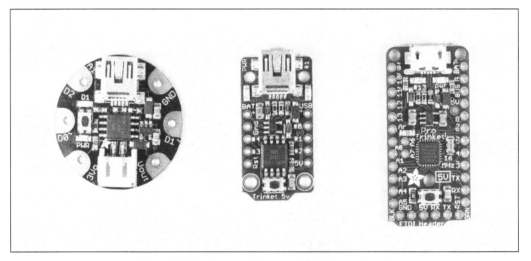

Figure 1.8: The Gemma, Trinket and Trinket pro board

In *Figure 1.8*, the first one from the left is the Gemma board. In the middle, you will find the Trinket board, and to the right, you have the Trinket pro board. Both the Gemma and Trinket board are based on the ATtiny85 microprocessor, which is a much smaller and cheaper processor, but comes with limitations. These boards only have three programmable pins, but what they lack in functionality, the make up for in size. The difference between the Gemma and Trinket board is the form factor, but the Trinket board also lacks a battery connector. The Trinket Pro board runs on an Atmega328 chip, which is the same chip used on the standard Arduino board to handle the USB communication.

This chip has 20 programmable pins, but also lacks a battery connector. The reason for using different types of boards in this book is that different projects require different functionalities, and in some cases, space for adding components will be limited. Don't worry though, since all of them can be programmed in the same way.

Connecting and testing your board

In order to make sure that you have installed your IDE correctly and to ensure your board is working, we need to connect it to your computer using a USB to USB micro cable, as show in *Figure 1.9*:

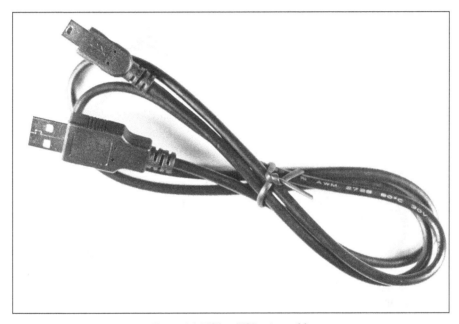

Figure 1.9: USB to USB micro cable

The small connector of the cable connects to your board, and the larger connector connects to your computer. As long as your board is connected to your computer, the USB port on the computer will power your board. In *Chapter 3, Bike Gloves*, we will take a closer look at how to power your board using batteries.

Once your board is connected to the computer, open up your IDE and enter the following code. Follow the basic structure of writing sketches:

1. First, declare your variables at the top of the sketch.
2. The setup you make is the first segment of code that runs when the board is powered up.

────────────────── **[11]** ──────────────────

3. Then, add the loop function, which is the second segment of the code that runs, and will keep on looping until the board is powered off:

```
int led = 7;

void setup()
{
pinMode(led, OUTPUT);
}

void loop()
{
digitalWrite(led, HIGH);
delay(1000);
digitalWrite(led, LOW);
delay(1000);
}
```

The first line of code declares pin number **7** as an integer and gives it the name LED. An integer is a data type, and declaring the variable using the name int allows you to store whole numbers in memory. On the FLORA board, there is a small on-board LED connected to the digital pin **7**. The next part is void setup(), which is one of the functions that always needs to be in your sketch in order for it to compile. All functions use curly brackets to indicate where the function starts and ends. The { bracket is used for the start, and } the bracket is used to indicated the end of the function. In void setup(), we have declared the mode of the pin we are using. All digital pins can be used as either an input or an output. An input is used for reading the state of anything connected to it, and output is used to control anything connected to the pin. In this case, we are using pin **7**, which is connected to the on-board LED. In order to control this pin we need declared it as an output.

If you are using a different board, remember to change the pin number in your code. On most other Arduino boards, the onboard LED is connected to pin **13**.

The void loop() function is where the magic happens. This is where we put the actual commands that operate the pins on the board. In the preceding code, the first thing we do is turn the led pin HIGH by using the digitalWrite() command. The digitalWrite() function is a built-in function that takes two parameters. The first is the number of the pin, in this case, we put in the variable led that has the value 7. The second parameter is the state of the pin, and we can use the HIGH or LOW shortcuts to turn the pin on or off, respectively.

Then, we make a pause in the program using the `delay()` command. The delay command takes one parameter, which is the number of milliseconds you want to pause your program for. After this, we use the same command as before to control the state of the pin, but this time we turn it `LOW`, which is the same as turning the pin off. Then we wait for an additional `1000` milliseconds. Once the sketch reaches the end of the loop function, the sketch will start over from the start of the same function and keep on looping until a new sketch is uploaded. The reset button is pressed on the FLORA board, or until the power is disconnected.

Now that we have the sketch ready, you can press the upload button. If everything goes as planned, the on-board LED should start to blink with a 1 second delay. The sketch you have uploaded will stay on the board even if the board is powered off, until you upload a new sketch that overwrites the old one. If you run into problems with uploading the code, remember to perform the following steps:

- Check your code for errors or misspelling
- Check your connections and USB cable
- Make sure you have the right board type selected
- Make sure your have the right USB port selected

Some notes on programming

Now that we know that your IDE and board are working, we will have a look at some more code. Programming with Arduino is a fairly straightforward process, but as with any other skill, it takes some practice. You should never feel stupid if you don't understand straightaway or if you can't get something to work as it should. This is a part of the process we call prototyping. When you prototype something, you may have a clear idea of what you want to do, but not a clear plan of how to achieve this. A big part of prototyping is the process of trial and error. If you try something and it does not work, then you try something different. A common misconception is that electronics break easily. It is true that components can break if connected the wrong way, but even breaking stuff can be helpful in the process of understanding how they work. However, it is very hard to break anything by code. Again, it is possible to break the microprocessor in most Arduino boards by uploading faulty code to them, but the IDE makes this nearly impossible since it always checks your code for errors before uploading it to the board. Microprocessors are logical in the strictest sense.

When learning to program, the most important part is to learn how to debug. Debugging is simply the process of finding where the problem is. Compile errors are the most obvious, since the IDE will let you know that your sketch contains an error somewhere. However, the IDE can only check for semantic errors and it does not know what you are trying to achieve. Your sketch might compile, but it still does not do what you want it to do. The deeper your understanding of the different commands, the faster you will become in the debugging process. In this book, I will explain the different commands as they are used in the chapters, but even if we use a lot of them, we will not cover all possible commands. If you want to learn more about all the possible commands, the Arduino website has a reference list in which you can find them (http://arduino.cc/en/Reference/HomePage). This book is aimed at readers who have some experience with programming for Arduino, and it does not include an introduction to programming. With this said, you should not feel excluded if you don't know how to program, since it should be possible to follow all the projects without a deeper understanding of programming. By following the instructions and code in this book, you should be able to create your own working version of all the projects. The following sketches are examples to get you going, which includes some of the basic functions and commands.

Note that all code that is proceeded by the use of // or /*......*/ are comments. The // (comment character) hides the line of code from the compiler, and will not be a part of the sketch. The /*......*/ (comment character) hides comments that are spread over multiple lines, where anything written in between /* and */ will be hidden from the compiler. It is good programming practice to add comments to document your code.

External LEDs and blinking

Now that we've tried a really simple example with the board by itself, it's time to add some extra components:

*   FLORA board
*   USB to USB micro cable
*   Two alligator clips
*   PCB mounted LED

Downloading the example code

You can download the example code files from your account at http://www.packtpub.com for all the Packt Publishing books you have purchased. If you purchased this book elsewhere, you can visit http://www.packtpub.com/support and register to have the files e-mailed directly to you.

In this sketch example, we will use an external LED, but if you want, you can stick with the on-board LED on the FLORA board. The LED used in following *Figure 1.10* is a special surface-mounted LED that is placed on a PCB. If you are using another LED, make sure to pair it with the right resistor. In the case of the custom LED found in *Figure 1.6*, the resistor is mounted on a PCB (printed circuit board). These LEDs are made with wearables in mind, and you can find them in most specialized electronic stores:

Figure 1.10. The LED connected to the board using alligator clips

To connect the LED to the board, we used alligator clips. Alligator clips are normal wires with metal clips at the end that are great for prototyping, and work especially well with wearable Arduino boards like the FLORA. LEDs have a positive and a negative side to them. In the case of the LED in *Figure 1.10*, these are marked on the PCB with a + sign for the positive and a - sign for the negative side. The positive side connects to pin **D12** on the FLORA board, and to complete the circuit, the negative side connects to **GND**.

Different speed blinking

The following sketches show how to blink the LED at different speeds, using the for loops:

```
int led = 12; //declares a variable called led connected to pin 12

void setup() {
  pinMode(led, OUTPUT); //declares led as an output pin
}

void loop() {
//start looping until the counter is bigger then 5
  for(int i=0; i<5; i++){
    digitalWrite(led, HIGH); //turn the led on
    delay(1000); //wait for a bit
    digitalWrite(led, LOW); //turn the led off
    delay(1000); //wait some more
  }
//start looping until the counter is bigger then 5
  for(int j=0; j<5; j++){
    digitalWrite(led, HIGH); //turn the led on
    delay(500); //wait for a bit
    digitalWrite(led, LOW); //turn the led off
    delay(500); //wait some more
  }
//start looping until the counter is bigger then 5
  for(int k=0; k<5; k++){
    digitalWrite(led, HIGH); //turn the led on
    delay(100); //wait for a bit
    digitalWrite(led, LOW); //turn the led off
    delay(100); //wait some more
  }
}
```

In this sketch, we use three different for loops. In the for loop, a counter, end condition, and counter increment is declared. All the counters in this sketch start on 0, and as long as the counter is smaller than 5, the for loop will keep on looping. For every time the for loop makes a loop, the counter is increased by one. Inside the for loop, the LED is first turned on and then there is a delay. The delay time is the difference between the three for loops.

Then, the LED is turned off and there is a new delay. Once the first loop has met the end condition, the second loop takes over, and so on. The for loops are good if you know you want to do a certain thing a certain number of times.

As you can see in the preceding sketch, the procedure for turning the LED on and off is the same for all of the `for` loops. This is a perfect opportunity to introduce a function into your code. Functions help save memory space. In the case of the sketch we are working with, memory space will not become a problem since the sketch is very small. However, as you progress, sketches will become bigger and bigger, and memory space may become a problem. Depending on the board you are using, there is a limit to how big you can make you sketches. Optimizing your code helps save space, but is also good coding practice, since it gives you a better overview of your code:

```
int led = 12;

void setup() {
  pinMode(led, OUTPUT);
}

void loop() {
  for(int i=0; i<5; i++){
    blinkLed(1000); //call the function and add the delay time
  }
  for(int j=0; j<5; j++){
    blinkLed(500); //call the function and add the delay time
  }
  for(int k=0; k<5; k++){
    blinkLed(100); //call the function and add the delay time
  }
}

/*declares the function blinkLed and adds a parameter that needs to be
included with the use of the function*/
void blinkLed(int delayTime){
digitalWrite(led, HIGH); //turn the led on
    delay(delayTime); //wait for a bit
    digitalWrite(led, LOW); //turn the led off
    delay(delayTime); //wait some more
}
```

The `blinkLed` function has been declared so it takes a parameter, which is `delayTime`. This variable is then used inside the function to set the speed of the blinking.

Summary

In this chapter, we have had a look at the different parts of the FLORA board and how to install the IDE. We also made some small sketches to work with the on-board LED. We made our first electronic circuit using an external LED.

In the next chapter, I will introduce you to some analog sensors that are suitable for working with wearable's. We will keep using LEDs as our output, to show how we can interact with the data gathered from the sensors, as well as how to control the intensity of the LED.

2
Working with Sensors

A sensor is a device that can detect changes or events and provide a corresponding output. The output is usually an electronic signal, for example, a **light dependent resistor (LDR)** outputs a voltage, which depends on the level of light cast on the sensors. When working with electronics, sensors are often divided into analog and digital sensors. Digital sensors can only detect two states, either on or off. The digital sensor can only distinguish if there is voltage going into the sensor or not. In code, this transfers into a 1 for voltage coming in and 0 if there is no voltage present. This is why they are called **digital** sensors, since they only operate in 0s and 1s. This means that these sensors only have two states, either on or off. A button, for example, is a digital sensor, which can only sense two states, if the button is pushed or not.

Analog sensors, however, can sense a range of values. The LDR, for example, is an analog sensor that changes the output voltage depending on the light level cast on the sensor surface. The problem with microprocessors is that they are digital by nature, and don't know how to handle analog information by default. This is why there are analog pins on almost all Arduino boards, which have an analog to digital conversion built in. These pins can read values ranging from 0 to 1023. In this chapter, we will have a look at some different sensors that may be useful for wearable projects and introduce them to readers that are not too familiar with programming yet.

In this chapter, we will take a look at a collection of analog sensors, which can be used to track movement and light. In the first two examples, we will focus on a stand-alone sensor component, which will involve building circuits using a breadboard. The remaining examples will use sensors that include prebuilt circuits on a PCB board. In this chapter, we will also take a look at different ways to communicate with our prototyping board and how to send data back to the computer.

> If you run out of digital pins, you can always use the analog pins even for reading digital sensors. Just remember that the output will be read in the range of **0** to **1023** and not **HIGH** and **LOW** if you are reading them as analog pins. The analog pins can be used as digital pins as well. Analog pin **A1** is the same as digital pin **14,** and so on.

Sensors

In this chapter, we will cover bend sensors, pressure sensors, light sensors, accelerometer, gyroscopes, and compass sensors. A sensor is a device that can detect events or changes of different kinds and can provide a corresponding output. It is a device that changes some characteristics due to external conditions and can be connected to a circuit, converting the signal so that it can be interpreted by a microprocessor.

With some sensors, you need to build you own circuitry and interpret the data provided by the sensor through code. Some complex sensors have a built-in communication protocol, which enables them to provide data corresponding to their function. The sensitivity of a sensor indicates how much the output data can change.

A bend sensor

The first sensor we will try out is a **bend sensor**. Sometimes it is also known as a **flex sensor,** and the name gives a good hint at what kind of sensor this is. As the name suggests, this sensor senses bends. The sensor works similarly to most analog sensors. These sensors are built on the same principal, where they take a voltage input and the sensor acts as a resistor that can change its own resistance. In *Figure 2.1* you will find all the components needed for this example. The component labeled as 1, in *Figure 2.1*, is a regular resistor, which measures 10kΩ. Resistors are some of the most common electronic components that you can find, and are used in almost all electronic devices. A resistor is used to limit the flow of current in an electronic signal. This means that if you apply a voltage at one end of the resistor, it will output a lower voltage at the other end. In the case of this example, we need a resistor to create what is known as a voltage divider in order to get good values from our sensor. The component labeled number 2, in *Figure 2.1*, is the bend sensor. These sensors detect bending in one direction. When left untouched, the bend sensor will measure about 10kΩ, and when fully bent it will measure around 20 kΩ. Number 3, in *Figure 2.1*, is the FLORA board we used in the previous chapter:

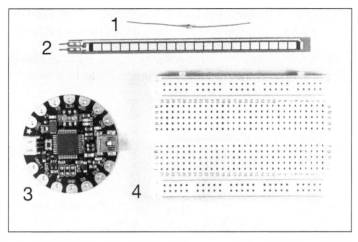

Figure 2.1: Showing 1. The 10kΩ resistor 2. A Bend/flex sensor 3. The FLORA board 4. The bread board

Number 4, in *Figure 2.1*, is a breadboard, and this is not technically a component. Breadboards are used for prototyping with electronics. When electronics are made, components are soldered together, but before you show how everything connects and what you want to make, it's good if you can try connections out before soldering them together. This is when a breadboard comes in handy. The breadboard in *Figure 2.1* is a miniature version of a standard breadboard. The lines of a breadboard are connected vertically in the middle of the board and horizontally at the edges. The + and − line marks the horizontal connections, while in the middle segment of the board the lines connect between the numbers and down from **a** to **j**. The following *Figure 2.2* shows an illustration of how the breadboard lines connect on the inside of the board:

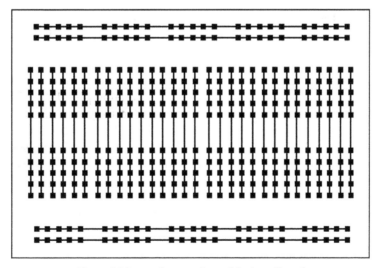

Figure 2.2: Internal connections of the breadboard

Breadboards make a circuit bulkier, but they are not used for the final designs. They are mainly used for trying out connections before soldering your components together.

Wires have also been added, so we can connect our sensor to the FLORA board later on. In *Figure 2.3*, you will find the necessary connections to be made:

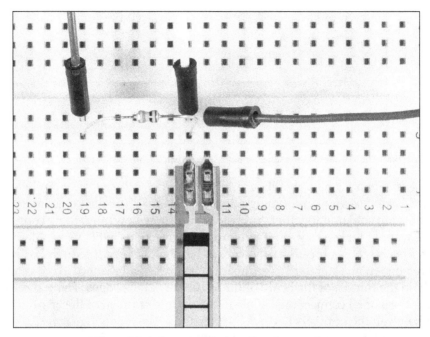

Figure 2.3: A close up of the breadboard connections

On the left side of the resister, we have a wire that connects to the **3.3 V** output of the FLORA board. Resistors have no polarity, which means the direction of connection does not matter. They work both ways so to say.

On the right side of the resistor, we connected one of the connections to the bend sensor, and one to a wire that we will connect to an analog pin of the FLORA board. The second connection of the bend sensor is connected to ground. In this circuit, power will be passed and limited through the resistor. Once on the other side of the resistor, some of the electricity will be passed through the sensor and some will be passed back to the analog pin. When we bend the sensor, we make it harder or easier for the electricity to pass through the sensor depending on how we bend it. The overflow of electricity that can't be passed through the sensor will be passed back along the analog pin, which we will read through our FLORA board.

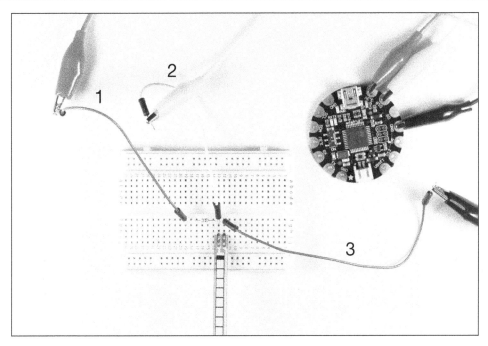

Figure 2.4: Connecting the bend sensor

The preceding *Figure 2.4* shows how to connect everything to the FLORA board using alligator clips:

- Cable number 1 connects to **3.3 V**
- Cable number 2 connects to **D10**
- Cable number 3 connects to **GND**

In order to check that everything works, we need to connect the FLORA board to our computer and write some code. For all of the sensor examples in this chapter, we will be using the serial library to send the data back to the serial monitor in your Arduino IDE. Serial communication is not used in every sketch, so to save memory space, the code that enables the communications is put in a library, which we can call upon when we need it. This is why all serial commands start with Serial., which means that we are calling the any command afterwards from the library. Serial communication is just one of the many communication protocols that can be used. Communication protocols are basically a set of rules on how two devices should act in order to communicate with each other.

```
//variable to store the analog pin
int bendSensor = 10;
```

```
void setup(){
//Start the serial communication
   Serial.begin(9600);
}

void loop(){
//Save the data from the sensor into storeData
   int storeData=analogRead(bendSensor);
//Print the data and add a new line
   Serial.println(storeData);
//Give the computer some time to receive the data
   delay(200);
}
```

The setup is where we initialize the serial communication. When this is done, you always need to declare the speed of your communication, which is calculated in baud, that is, same as bits per second. The speeds are fixed, so if you want to see what speeds are available, open up your serial monitor and you should find something that looks like *Figure 2.5*:

 When using serial communication, always use a delay in your sketch to give the receiving side some time to be able to read it. If you don't need to read it at human readable speed, a delay of about 30 milliseconds will suffice.

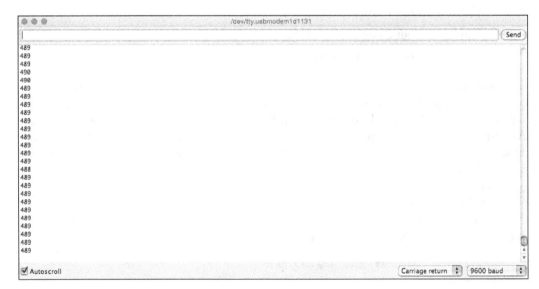

Figure 2.5: Showing the serial monitor

In the bottom of the right corner in the serial monitor you will find a drop-down menu with the available speeds. If there is no reason for choosing a particular speed, I usually use **9600 baud**. This rate is fast enough for most applications and is supported by most devices. At the top of the serial monitor, you will find an input box where you can send information to the FLORA board. The preceding *Figure 2.5* also shows data that is sent from the sensor. In my case, the sensor outputs a steady value of 489 when left untouched. When the sensor is fully bent, the value increases to around 650. In the previous code example, the data was sent back to the computer using the Serial.println() command, which adds a carriage return. This means that when the data is received by the computer, it adds it to a new line when presented on the screen. You can send data without the carriage return by simply using the Serial.print() command. However, this will keep on printing data on the right, on the same line, until the Arduino sends a line return.

The data can instead be formatted by the serial monitor using the drop-down menu to the left of the speed menu. In this drop down, you will find different formatting options such as adding a new line, carriage return, both, or none of them. Remember that the built-in formatting in the serial monitor only works for incoming data. On outgoing data, you will need to format your data in code.

In the next example, we will take a look at how to interact with a pressure sensor.

The pressure sensor

This example follows the same principals as the previous bend sensor, but this time, we will try out a pressure sensor. The pressure sensor does the same thing as the bend sensor where it regulates a voltage output depending on the pressure applied to the sensor. If the sensor is pressed, it will add to the resistance and the voltage coming out the other end of the sensor will be lower. Again, we will use the FLORA board to detect the changes in the value of the output.

You will need the following components found in *Figure 2.6*:

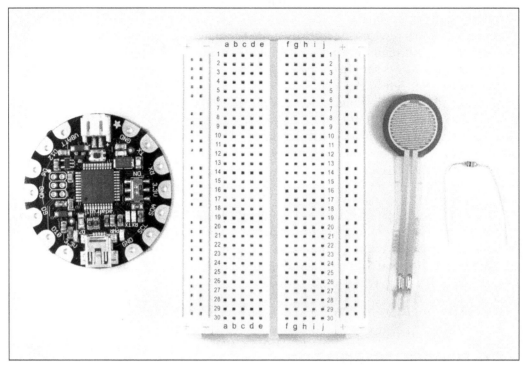

Figure 2.6: The FLORA board, bread board, pressure sensor and 10 kΩ resistor

From the left, you have the FLORA board, breadboard, pressure sensor, and a 10 kΩ resistor. In order to make the circuit, we will be using the same setup as we did in the bend sensor example. The pressure sensor is also known as a force sensor and can be used to detect physical force, touching, or even weight. If weight is what you are looking for, this might not be the best choice since this sensor doesn't output an actual weight value, though by using some fixed weights and some calibration you could make a crude scale.

In *Figure 2.7* you will find the necessary connections that need to be made:

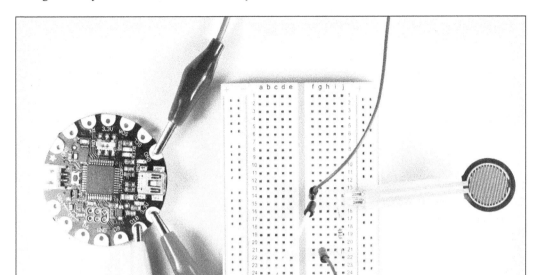

Figure 2.7: Showing the connections to the FLORA board

Let's take a closer look at the connections:

- On the side of the resistor with only one wire, we have a connection to the **3.3V** pin on the FLORA board.
- On the other side of the resistor, we have a connection to the first pin of the pressure sensor and a wire connection to **D12** on the FLORA board.
- The second pin of the pressure sensor connects to **GND** of the FLORA board.

To test the range of the sensor, you can use the same code used for the bend sensor, as it prints the analog data back to your serial monitor. In the case of the sensor, I was using, it gave me almost the full range from 0 up to 1023. In the following example, we use the value from the sensor to make a small visualization using the onboard LED on the FLORA board:

```
//declare variables to store pin numbers
int pressureSensor=12;
int led=13;
void setup() {
  //declar pin as an output
  pinMode(led,OUTPUT);
}

void loop() {
  // read the input on analog pin 12:
  int pressureValue = analogRead(pressureSensor);
  //turn the led on
  digitalWrite(led,HIGH);
  //wait using the value from the sensor
  delay(pressureValue);
  //turn the led off
  digitalWrite(led,LOW);
  //turn the led on
  delay(pressureValue);
}
```

In the sketch, we use the data from the sensor to set the delay time of the LED. When you press the sensor, this will lower the value, which will lessen the delay between blinks. This in turn will mean that the LED will start to blink faster. Remember that your sensor might have a slightly different value range, so you might need to modify the value in order to find a satisfactory blinking speed. The best way is to print out your sensor data to the serial monitor using serial communication in order to check your range of values. Then, you can add or subtract with any amount you prefer inside the delay, an example of this is shown as follows:

```
digitalWrite(led,HIGH);
  //wait using the value from the sensor
  delay(pressureValue+1000);
  //turn the led off
  digitalWrite(led,LOW);
  //turn the led on
  delay(pressureValue+1000);
```

In the next sketch, for example, we will make a mini game where the goal is to find the sweet spot. The code is set up so that you have to press the sensor in order to find the hidden spot defined in the code. Once you find it, the LED will start to shine for as long as you can hold the spot:

```
//declare variables to store pin number
int pressureSensor=10;
int led=7;

void setup() {
  //declare pin as an output
  pinMode(led,OUTPUT);
  Serial.begin(9600);
}

void loop() {
  //read the input on analog pin 12:
  int pressureValue = analogRead(pressureSensor);
  //as long as the value is in between 500 and 700 keep on looping
  while(pressureValue>500 && pressureValue<700){
    //turn the led on
    digitalWrite(led,HIGH);
    //check so that the value
    pressureValue = analogRead(pressureSensor);
  }
  //turn the led off
  digitalWrite(led,LOW);
}
```

The sweet spot is hidden in the range of 500 to 700. When the value is in the sweet spot range, we enter the while loop, which will keep on looping until the condition for the while loop is no longer met. This is why we need to add an additional reading of the sensor inside the loop, or the code will get stuck inside the while loop forever. As long as you can keep the value inside the range, the LED will light up. If you press the sensor too hard or too little, the while loop will be broken and the LED will turn off. If you find it too easy to find the sweet spot, try to decrease the range. This will make it much hard to find the spot.

Light dependent resistors

In this example, we will have a closer look at an LDR. The principal behind it is the same as the bend sensor used in previous bend sensor example. Depending on the light levels cast onto the sensor, the sensor changes its output voltage. LDR comes in different shapes and sizes and the LDR used in *Figure 2.8* comes premounted with a surface mounted resistor on a small PCB manufactured by Adafruit. The principal of the PCB is similar to the circuit created using a breadboard and components in *Figure 2.3*.

Figure 2.8: A PCB mounted LDR and resistor

The difference is that in the case of the PCB, in this example, everything is mounted nicely and we only need to attach it to the FLORA board using alligator clips. As shown in *Figure 2.8*, the **LDR PCB** has three connections **VCC**, **OUT**, and **GND**. Sometimes **VCC** is used to indicate power in. **OUT** in this case is the output signal, which is an analog signal. In *Figure 2.9*, you will find the necessary connections to the FLORA board. There is not a lot of space between the connections, so make sure that the alligator clips do not touch one another.

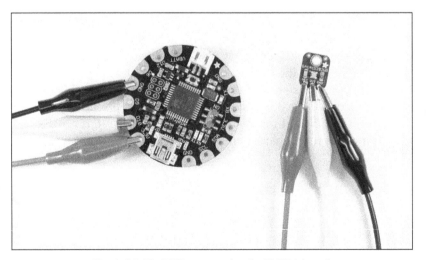

Figure 2.9: The LDR connected to the FLORA board

Since we are using the same connections as in the previous example in the chapter, you can use the same code to test you sensor. The following sketch is based on the same code as used in the bend/flex sensor example, but we have implemented some code to remap the range of values from the sensor into a new larger range of values.

```
//Variable to store the pin
int ldrSensor = 10;

void setup(){
//Start the serial communication
  Serial.begin(9600);
}

void loop(){
//Save the data from the sensor into storeData
  int storeData=analogRead(ldrSensor);
//Re-map storeData to a new range of values
  int mapValue=map(storeData,130,430,0,2000);
//Print the re-mapped value
  Serial.println(mapValue);
//Give the computer some time to print
  delay(200);
}
```

The map() function in the code is used to take on a value range and remap it to a new range of values. In order for this function to work, you need to know the range of values possible from the senor you are using. Using the code from the bend sensor example, you can do this. In order to check you LDR sensor range, you need to print the data to your serial monitor. First, check the value you get when the LDR is not covered and write this down. Then, cover the LDR as much as possible and check the value. Make sure to remember this value as well since you'll need it.

The map function takes five parameters map(sensorValue, sensorMin, sensorMax, desiredMin, and desiredMax). The first parameter is the actual sensor data. The second and third are the minimum and maximum values from the sensor, which you wrote down from the previous test. The fourth and fifth are the minimum and maximum values of your desired range. In the case of the LDR that I an using, this gives a range of values between 130 to 430, which we will choose to remap into a value range of 0 to 2000.

The map function comes in handy in many cases when you want to use the data from a sensor but the default range does not fit your needs. Using the map function, you can translate it into any desired range.

The accelerometer, compass, and gyroscope

In the following example, we will try out an accelerometer, which is a device used for sensing g-force. The accelerometer can sense movement in different directions. Normally, there are two types of accelerometer: the 2 axes and 3 axes. The 2 axes measures in two directions, left to right and front to back, and these directions are often named as the x axis and y axis. The 3 axes accelerometer also measures in a z axis, which is up and down. The accelerometer measures movement relative to its own position, which means that when you move it, it measures the g-force in the direction you are moving it.

The sensor used in this example is the FLORA accelerometer/compass/gyroscope, which is actually three sensors in one. Besides the accelerometer, it holds a compass, known as a **magnetometer**, which senses where the strongest magnetic field is coming from. If there are no other magnets near by, the strongest magnetic field comes from the Earth's north pole.

The third sensor is the gyroscope, which is a device that can sense rate of orientation on a spin axis through external torque. This basically means that the gyroscope can sense in which direction you are turning, compared to the accelerometer that senses which direction you are tilting the sensor.

In previous examples, we have been using serial communication to communicate with our sensors, but in the case of the FLORA accelerometer/compass/gyro, we will be using a different communication protocol called **I2C**. The **I2C** is a protocol developed by Philips, and is used to connect low-speed devices to microprocessors, among other things. It uses a master-slave system, which involves one device acting as the master and all the other devices connected to the communication channel acting as slaves. The master can talk to all of the slaves, but the slaves can only talk to the master. **I2C** is a two-wire protocol where you can hook up multiple slaves to the same communication lines in any order you want, since all slaves are given a unique ID to keep track of them. Keeping track of the communication is the master job so no more than one slave can communicate at the time. In the case of this example, we only have two devices, the FLORA, which will be acting as a master, and the accelerometer/compass/gyro sensor, acting as slaves. Using **I2C**, we can communicate with all three sensors without connecting them on separate communication channels.

The FLORA accelerometer/compass/gyro is based on a chip called **LSM9DS0**, which by default is a bit tricky to communicate with. There are many breakout boards based on the LSM9DS0 available, and most of them are supplied with Arduino libraries, even the FLORA one. On the following GitHub link, `https://github.com/adafruit/Adafruit_LSM9DS0_Library` you can find the Adafruit LSM9DS0 library written by Kevin Towsend, which is an excellent library that simplifies the communication with the LSM9DS0. Just press the **Download ZIP** button on the GitHub repository and the library will start downloading.

As I mentioned before, libraries contain code that is usually too long and complicated to copy. In most cases, the code is used for different functionalities, and does not need to be changed. In order to make life easier for others to use their code, some kind souls have made them into libraries so that they can easily be shared.

Once the LSM9DS0 library is downloaded, you need to unzip the folder and rename it to `Adafruit_LSM9DS0`. then place it into your libraries folder, inside you Arduino folder, by navigating to **Arduino | libraries**. If this is the first time you are installing a library, this folder might be missing, and in that case, you just create a folder inside the Arduino folder called `libraries`. To find out where your libraries folder is located, you can check in the Arduino IDE by navigating to **Arduino | preferences**.

This will open up a new window that shows the full path to your libraries folder.

The LSM9DS0 library is dependent on another library by Adafruit, which you also need to install, and you can find it on the following link:

`https://github.com/adafruit/Adafruit_Sensor`.

Install it as you did with the LSM9DS0 library and rename it as Adafruit_Sensor. When the libraries are installing, we can carry on with hooking up our sensor to the FLORA board, as shown in *Figure 2.10*:

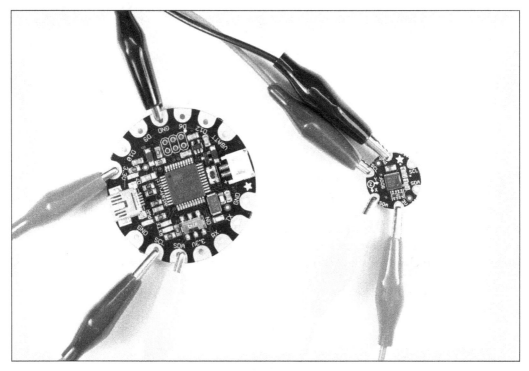

Figure 2.10: Connecting the sensor to the FLORA board

The first connection that needs to be made is the **3.3 V**, which connects to the **3.3 V** pin on the FLORA board. Then connect **SDA**, **SCL**, and **GND** to the same pins with the same name on the FLORA board. Once all the connections have been made, you can enter the following code into the IDE and upload it to the FLORA board:

```
/*Import the necessary libraries. Wire.h and SPI.h are included in the
IDE and does not need to be downloaded*/
#include <Wire.h>
#include <SPI.h>
#include <Adafruit_LSM9DS0.h>
#include <Adafruit_Sensor.h>

// set up i2c
Adafruit_LSM9DS0 lsm = Adafruit_LSM9DS0();

void setupSensor()
```

```
{
  /*Set the accelerometer range where the range can be changed to
4G,8G or 16G */
  lsm.setupAccel(lsm.LSM9DS0_ACCELRANGE_2G);
  //lsm.setupAccel(lsm.LSM9DS0_ACCELRANGE_4G);
  //lsm.setupAccel(lsm.LSM9DS0_ACCELRANGE_6G);
  //lsm.setupAccel(lsm.LSM9DS0_ACCELRANGE_8G);
  //lsm.setupAccel(lsm.LSM9DS0_ACCELRANGE_16G);
}

void setup()
{
  //Wait for communication to start
  while (!Serial);
  //Set the baud rate
  Serial.begin(9600);
  //Print message to serial monitor
  Serial.println("Starting communication");

  // Try to initialize and warn if we couldn't detect the chip
  if (!lsm.begin())
  {
    Serial.println("Something went wrong, check you connections");
    while (1);
  }
  Serial.println("Connection established");
  Serial.println("");
  Serial.println("");
}

void loop()
{
  //Get data from the sensor
  lsm.read();
  //Print the accelrometer sensor data to the serial monitor
Serial.print("Accel X: ");
Serial.print((int)lsm.accelData.x);
Serial.print(" ");
Serial.print("Y: ");
Serial.print((int)lsm.accelData.y);
Serial.print(" ");
Serial.print("Z: ");
Serial.println((int)lsm.accelData.z);
Serial.print(" ");
  delay(200);
}
```

This code example initializes the accelerometer and prints the data from the sensor back to the serial monitor. Don't forget to open the monitor in order to see the data flow. If the sensor is left untouched, you will find that the sensor data does not change that much. In the case of my sensor, the values on all three axes ranged from +30 to -30 as some noise is expected. When I start tilting the sensor to the left, the value starts to increase on the *x* axis, and if I tilt it to the right, the *x* axis value starts to decrease. The value range on your sensor might be different, so try it out yourself by tilting the sensor in different directions while looking at the serial monitor to get a sense of you sensor range. Accelerometers can only sense movement in 180 degrees, so once your flip it to 180 degrees, the value starts over for the corresponding value on the other side.

In the next code example, we added the functionality of both the compass and gyroscope along side the accelerometer code in order to show the sensors full functionality.

```
//Import the necessary libraries
#include <Wire.h>
#include <SPI.h>
#include <Adafruit_LSM9DS0.h>
#include <Adafruit_Sensor.h>

// set up i2c
Adafruit_LSM9DS0 lsm = Adafruit_LSM9DS0();

void setupSensor()
{
    /*Set the accelerometer range where the range can be changed to
4G,8G or 16G*/
    lsm.setupAccel(lsm.LSM9DS0_ACCELRANGE_2G);


    //Set the magnetometer sensitivity where the range can be changed to
4GAUSS, 8GAUSS or 12GAUSS
    lsm.setupMag(lsm.LSM9DS0_MAGGAIN_2GAUSS);


    /*Setup the gyroscope sensitivity where the range can be changed
500DPS or 2000DPS*/
    lsm.setupGyro(lsm.LSM9DS0_GYROSCALE_245DPS);

}

void setup()
{
   //Wait for communication to start
   while (!Serial);
```

```
  //Set the baud rate
  Serial.begin(9600);
  //Print message to serial monitor
  Serial.println("Starting communication");

  // Try to initialize and warn if we couldn't detect the chip
  if (!lsm.begin())
  {
    Serial.println("Something went wrong, check you connections");
    while (1);
  }
  Serial.println("Connection established");
  Serial.println("");
  Serial.println("");
}

void loop()
{
  //Get data from the sensors
  lsm.read();
  //Print the accelrometer sensor data to the serial monitor
  Serial.print("Accel X: "); Serial.print((int)lsm.accelData.x);
Serial.print(" ");
  Serial.print("Y: "); Serial.print((int)lsm.accelData.y);
Serial.print(" ");
  Serial.print("Z: "); Serial.println((int)lsm.accelData.z);
Serial.print(" ");
  //Print the compass sensor data to the serial monitor
  Serial.print("Compass X: "); Serial.print((int)lsm.magData.x);
Serial.print(" ");
  Serial.print("Y: "); Serial.print((int)lsm.magData.y);
Serial.print(" ");
  Serial.print("Z: "); Serial.println((int)lsm.magData.z);
Serial.print(" ");
  //Print the gyro sensor data to the serial monitor
  Serial.print("Gyro X: "); Serial.print((int)lsm.gyroData.x);
Serial.print(" ");
  Serial.print("Y: "); Serial.print((int)lsm.gyroData.y);
Serial.print(" ");
  Serial.print("Z: "); Serial.println((int)lsm.gyroData.z);
Serial.println(" ");
  //wait for a bit
  delay(200);
}
```

With the help of the preceding code, we will now be able to show the sensors full functionality.

Summary

In this chapter, we had a look at some different analog sensors that might be suitable for working with wearables. What defines an analog sensor is that they can output a range of values, not just 0s and 1s. The selection of sensors used in this chapter is just a small portion of the sensors available on the market, and if you can think of something that you might want to sense in the future, there is a good chance there is a sensor available for it.

The purpose of the chapter has also been to introduce readers not familiar with Arduino programming to some new commands, as well as two different communication protocols for interfacing with more complex sensors. Sensors that use some form of communication protocol are often referred to as complex sensors, and the ones that don't, such as the bend sensor and LDR used in this chapter, are normally just called sensors.

Finding the right sensor for your project may be tricky sometimes, and some of the sensors available can be expensive. Think twice before you pick your sensor. Sometimes it is possible to use cheaper components for the same purpose. For example, if you need to track movement very precisely, a combination sensor such as the accelerometer/compass/gyro in this chapter is great. But if its basic movement or non-movement you are looking for, you might be fine with using a inexpensive tilt sensor, which acts as a button with a metal ball inside it. When the tilt sensor moves, the metal ball rolls over and completes the circuit. This can be used as a digital switch that switches between HIGH and LOW, instead of using one with a large number of continuous variables, which need processing. This simpler option might save you time and money.

All of the sensors in this chapter were picked for their "wearability" in terms of function. They were also picked because they are fairly easy to get started with. When it comes to developing wearable project, you will find that these sensors come in handy. When it comes to the human body, we have a tendency to move around, and this is where the accelerometer/gyro/compass sensor may come in handy. The bend/flex sensor might come in handy if you want to track more specific movements of the body, especially if placed in the right location such as the cress of an arm or knee. Even if LDRs are simple sensors, they are good enough to distinguish if some one is inside or outside.

As I just mentioned, this chapter and the first one are designed to get you going if you are completely new to working with electronics and programming. In the next chapter, you will create your first project, which implements both analog sensors and LEDs in order, to create two automated turning light bike gloves.

3
Bike Gloves

In this chapter, we will take a look at how we can create an automated lighting system for indicating turns while biking. When biking in the dark, it can sometimes be hard to see when a cyclist indicates to make a turn or stop. This project will help in these situations by adding some light to the bicyclist's hands. The goal is to create a set of devices that work both with and without gloves, which can be worn on top of your hands. Since we have the advantage of using microcontrollers, we will also build these bike gloves so that they can detect if lighting is needed or not.

In this chapter, you will learn more about accelerometers, light sensors, and some more advanced coding. The chapter will also introduce the basic concept of gesture recognition and how to program for interactivity. We will also take a look at how to finalize you project and power the device using external power sources.

Electronics needed

For this project, we will need the following components as shown in *Figure 3.1*. Up on the right, you will find the FLORA board used in previous chapters. The round sensor at the bottom in *Figure 3.1* may seem familiar. This is the same accelerometer/compass/gyro we used in the previous chapter. We will be using this sensor to track some basic gestures, so we know when to turn on the light and if we want to make a turn or stop. To the left and right of the accelerometer/compass/gyro you will find a set of white and red LEDs. These are surface mounted LEDs specially made for working with wearables. For this project, you can use any amount of LEDs and any color you want, but the idea is to have white LEDs for indicating direction and red ones for indicating stopping. When choosing the color of your LEDs, first make sure that they follow the biking laws in your country. Some countries have special rules for the color of the light in relation to their position on the bike. For instance, in some countries, you're only allowed to use green lights on the front of the bike and red ones on the back.

The last component in the top-left corner is a light sensor based on the **TSL2561** chip. This sensor senses luminosity and is an advanced digital sensor:

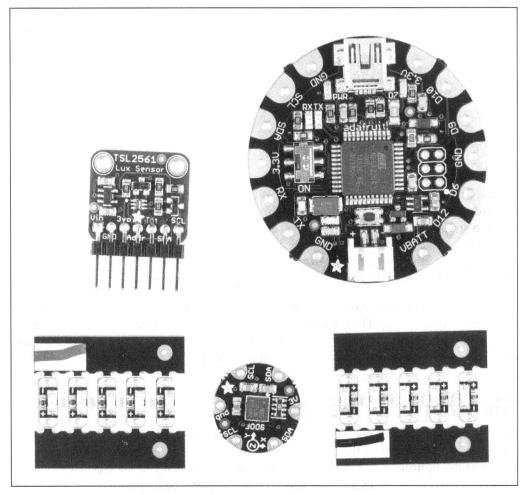

Figure 3.1: The FLORA board, LEDs, accelerometer/compass/gyro, TSL2561

Light is usually measured in the unit candela, while lumens measure the total amount of light. Lux is a unit defined by lumens per area, where 1 lux is the same as 1 lumen per square meter. The TSL2561 sensor can measure anything from -0.1 to 40000+ lux, which makes it a pretty good sensor for detecting light levels.

You might now wonder why we are not using the light sensor from previous examples. The reason is that the analog light sensor used in the previous chapter is not very precise and will need to be constantly calibrated, since riding a bike might entail moving around in multiple light conditions. The TSL2561 sensor actual reads the level of light, making it more reliable and providing more interesting scenarios for use. Other components needed for this chapter are:

- 220 Ω resistors
- Straps or wristband with an approximate width of 4 cm
- Soldering iron
- Wires
- Pliers and scissors

Trying out the TSL2561

Since the TSL2561 sensor is an advanced digital sensor, we will need to communicate with over I2C, since we can't read it as an analog sensor. Unlike other simpler light sensors, the TSL2561 chip measures both infrared and visible light to better approximate the response of our own eyes. As many of the components used in this book, the TSL2561 breakout board we will be using is from Adafruit, which also supplies a library for this sensor. This makes communicating with the sensor a bit easier. You can find the library on their GitHub: `https://github.com/adafruit/Adafruit_TSL2561`.

Download and install the library into you Arduino library folder. If you don't remember how to do this or if it is your first time installing a library, please take a look at *Chapter 2, Working with Sensors* where we explained the processes.

Once you have your library installed, its time to connect the TSL2561 sensor to you FLORA board. For testing the TSL2561 sensor, we will use alligator clip wires to connect the FLORA board. Remember there is not a whole lot of space in between the pins, so makes sure that the alligator clips do not connect to any other pins. *Figure 3.2* shows the necessary connections that need to be made to the FLORA Board:

Figure 3.2: Showing the connections from the TSL2561 to the FLORA board

The **3.3 V** pin on the TSL2561 board connects to the **3.3 V** pin on the FLORA board. The **SDA** pin on the TSL2561 board connects to the **SDA** on the **FLORA**, and the **SCL** pin connects to the **SCL** pin. Don't forget to connect **GND** on both boards. As always, in order to be complete a circuit, ground needs to be shared in-between components.

Once everything is connected, we can upload the following code to the FLORA board to check if we can receive any data:

```
//Include the libraries needed
#include <Wire.h>
/*If you a missing the Adafruit sensor library have a look at chapter
2 for a link to the library and install instructions*/
#include <Adafruit_Sensor.h>
#include <Adafruit_TSL2561_U.h>
```

```
//Give the sensor a ID which in this case is 12345
Adafruit_TSL2561_Unified tsl = Adafruit_TSL2561_Unified(TSL2561_ADDR_
FLOAT, 12345);

//Function to configure the sensor
void configureSensor()
{
  /* This set the gain to auto, you can also manual set the gain to no
gain or 16x the gain.*/
tsl.enableAutoRange(true);
  // To manually set the gain use the following commans
  // tsl.setGain(TSL2561_GAIN_1X);
  // tsl.setGain(TSL2561_GAIN_16X);

  /* modifying the integration time gives you better a resolution on
you readings (402ms = 16-bit data) */

/*Faster reading but lower resolution
tsl.setIntegrationTime(TSL2561_INTEGRATIONTIME_13MS);
Medium speed and medium resolution tsl.setIntegrationTime(TSL2561_
INTEGRATIONTIME_101MS);
High resolution but slow speed which we will be using for this
example*/
  tsl.setIntegrationTime(TSL2561_INTEGRATIONTIME_402MS);

}

void setup()
{
//Start serial communication
  Serial.begin(9600);
//Print test message
  Serial.println("Light Sensor Test");

  /* Initialize the sensor */
  if(!tsl.begin())
  {
    /* There was a problem detecting the sensor check your wires */
    Serial.print("Cant detect the TSL2561 detected Check your
wiring");
    while(1);
  }
```

```
    /* Setup the sensor gain and integration time */
    configureSensor();


}

void loop()
{
    /* Get a new sensor event */
    sensors_event_t event;
    tsl.getEvent(&event);

    /* If there is light display the results */
    if (event.light)
    {
      Serial.print(event.light); Serial.println(" lux");
    }
    else
    {
      /* If there is no light  */
      Serial.println("Cant see anything..Show me some light");
    }
    delay(250);
}
```

Note that you can set the gain of this sensor. Gain is the measurement of the ability to increase the power or amplitude of a signal. What this means is that you can set the sensitivity of a sensor by increasing or decreasing the gain. The TSL2561 can also set the gain automatically using the command:

```
tsl.enableAutoRange(true);
```

In order to get a sense of how the sensor reacts under different light conditions, I recommend you do some testing with the sensor outdoors.

Detecting gestures

In this next part, we will have a look at how to detect gestures with the accelerometer. For this project, we will not be using the gyro and compass. However, remember that even though this chapter will eventually end, it does not mean your projects have to. Once you are done, you can keep developing the project further and implementing the compass into the bike gloves, which may come in handy in the future.

First, let's take a look at how we can detect gestures. We will need three different interactions, the holding of the handlebar, indicating a turn, and stopping. We need to know that we are holding the handlebar, since this will act as our default state when the light should be off. In order to turn the white LEDs on, we need to know when the hand is lifted to indicate a turn, and we need to know when the hand is in a vertical position so we can turn on the red LEDs to indicate that we are stopping. Take a look at *Chapter 2, Working with Sensors* in order to connect the accelerometer. As mentioned earlier, the connections that need to be made from the accelerometer to the FLORA board is **3 V** to **3.3 V, GND** to **GND, SDA** to **SDA,** and **SCL** to **SCL**:

```
#include <Wire.h>
#include <SPI.h>
#include <Adafruit_LSM9DS0.h>
#include <Adafruit_Sensor.h>

// i2c
Adafruit_LSM9DS0 lsm = Adafruit_LSM9DS0();

void setupSensor()
{
  //Set range of the accelerometer
  lsm.setupAccel(lsm.LSM9DS0_ACCELRANGE_2G);
}

void setup()
{
  //wait until serial starts
  while (!Serial);

  Serial.begin(9600);

  // Try to initialize sensor and warn if it cant find it
  if (!lsm.begin())
  {
    Serial.println("Cant find the sensor, Check your connections!");
    while (1);
  }
  Serial.println("Found the sensor");
  Serial.println("");
  Serial.println("");
}

void loop()
```

```
  {
    //read the sensor
    lsm.read();
    if(lsm.accelData.x<0 && lsm.accelData.y<0){
      Serial.println("Holding handlebar");
    }
    if(lsm.accelData.y>1000 && lsm.accelData.x<0){
      Serial.println("Stopping");
    }
    if(lsm.accelData.x<-1000 && lsm.accelData.y<3000){
      Serial.println("Turning");
    }
  delay(500);
  }
```

In the sketch, we are using a combination of the x and y axis to detect the gestures. Note that the values used in the preceding sketch were generated from the direction of my sensor, and the reference value that you need might be something completely different. It all depends in which direction you sensor is placed. The easiest way to figure out the gestures is to place the sensor flat on the table and read the x and y values. This will act as your default state, or the "holding of the handlebar" state. In my case, I got steady negative readings, so any changes to the positive, or bigger then 0 as presented in the code, would indicate that the hand is moving. Hold the sensor vertically in order to figure out what value readings will be suitable for indicating a stopping gesture. From the stopping position, turn the sensor about 45 degrees and you will find the turning position.

This sketch is just a rough estimate and shows how you can make your own simple gesture recognition in code. The values we base our gestures on also depend on whether we are calibrating the left or right glove. In the final version of the code for this project, we will need to modify the values. In order to fine-tune everything, we need to put all the components together.

Making a glove

For this first project, we will keep the casing of the project fairly simple. When it comes to your own casing, you can use whatever material you find suitable and let your creativity flow. The idea for bike gloves lights in this chapter is to encase the electronics, which can then be attached to a pair of gloves using Velcro. The reason for this is to be able to swap for a wristband during warmer weather.

In *Figure 3.3* you will find the connection layout of how to connect all the LEDs, the TSL2561, and the accelerometer to the FLORA board:

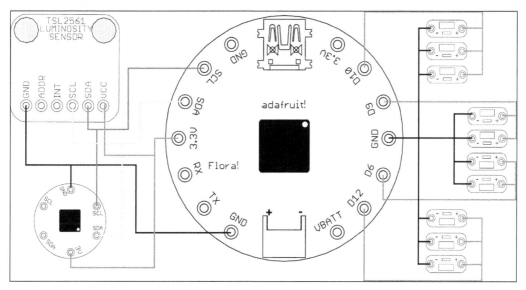

Figure 3.3: The connection needed for all the components

The LEDs are connected in parallel, in combinations of three and two LEDs. There are many ways you can connect electronic components, and the two easiest are parallel and series connection. When connected in parallel, the same voltage is applied to all the components. When connected in series, the same current is passed through all the components. The reason for connecting more than one LED to the pins is so that we can control multiple LEDs with one pin. The reason for having them in pairs of three and two LEDs is to make a small animation when turning and stopping. The setup in this chapter is just an example of the LED configuration, and you can use any setup you like for your project. Remember that the special surface mounted LEDs in this chapter have an on-board 220 Ω resistor built in. If you are using a normal resistor, you need to add an external resistor to your circuit. As you can see, both sensors are connected to the same communication pins on the FLORA board. While using I2C, you can connect multiple devices in the same way and have them communicate with the microcontroller. One disadvantage with I2C is that we can only communicate with one device at a time because it uses the same wire for all devices on the I2C port. However, this is not a problem, since the Atmega chip used on most Arduino boards is pretty fast. Switching the communication from one device to the other is how it is done, and in terms of human interaction, the FLORA board is much faster, and that is why we can precise.

To start we will solder wires to the LEDs and mount them on the casing. For the casing I'm using a transparent spray can cap, which I have cut down a bit to make it narrower. It has a nice round form factor and will protect the circuit from rain if case you get stuck in bad weather. The spray can cap I found is slightly bigger than the FLORA board, which is great since we need to fit in all the wiring and the rest of the components. Stack all the components together to give a rough estimate of how much you need to cut from the cap.

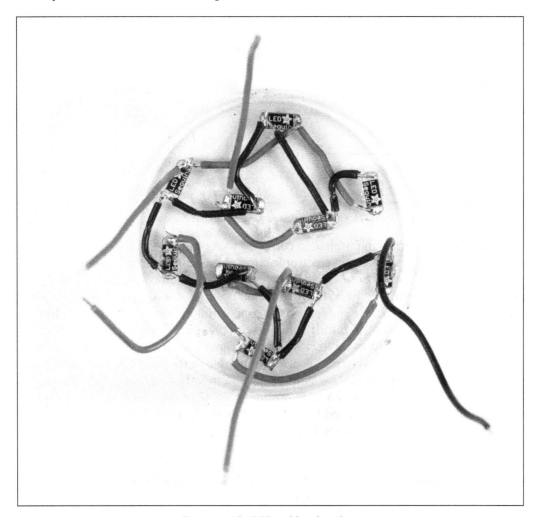

Figure 3.4: The LEDs soldered in place

Figure 3.4 shows the LEDs inside the can cap. In order to fix them to the cap, use hot glue and glue them with the LEDs facing down. The idea is that the LEDs will shine through the transparent cap. You can use any glue you want, but hot glue is great, since if you place something in the wrong place, you can still remove it and try again. The LEDs in *Figure 3.4* are soldered together according to schematic in *Figure 3.3*. All the ground connections have been soldered together since ground is the same for all components. From each set of LEDs and the ground connection we have added a wire so that we can connect them to the FLORA board. The next step is to connect the TSL2561 and accelerometer to the FLORA board.

Figure 3.5: The TSL2561 and accelerometer connected and soldered to the FLORA board

Figure 3.5 shows how the sensors can be connected to the FLORA board. The setup to the left in *Figure 3.5* shows how the sensors have been attached to the FLORA board using hot glue. The hot glue also acts as a separator, so the sensors do not come in direct contact with the FLORA board, avoiding short circuits. To the left in *Figure 3.5*, you will find the necessary connections soldered. If you're new to soldering, I would suggest training before you attempt soldering the sensor to the board, since it is a bit tricky. Beware of cold soldering and exposed wiring, and take your time with the soldering. Cold soldering are solder joints with a bad connection due to the solder not melting properly. In the following link, you can find more information about cold soldering and what it looks like: https://learn.adafruit.com/adafruit-guide-excellent-soldering/common-problems.

The FLORA board is actually designed to be stitched into garments using conductive thread, but when it comes to small project such as this one, the increased connection pads really make things a bit easier to solder to the FLORA board.

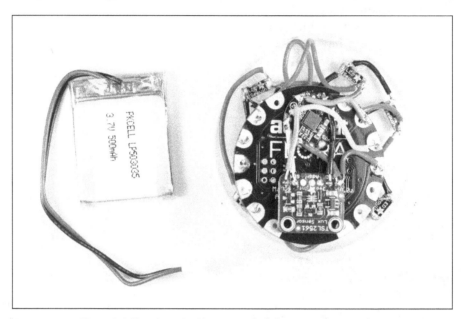

Figure 3.6: Showing a 3.7V battery and all the components in place

Once you have soldered the sensors to the FLORA board it is time to connect the LEDs and place all the components inside the cap. The LEDs connect to available digital pins left on the FLORA board. You can have a look at the final code for this project in order to see which LEDs connect to which pins. It is no problem if you end up soldering the wrong LEDs to the wrong pins, since you can always swap the order of the pins in the code later on. In *Figure 3.6*, you will find all the components in place, and the battery we will be using for this project. I have chosen a 3.7 V 500mAh battery for this project. The primary reason for choosing this battery is that it fits inside the cap and it outputs the right voltage. The 500 mAh should also keep the light going for some time before you need to recharge the battery. Batteries are measured in voltage and amperes. The abbreviated unit mAh is short for milliampere hours, which will determine how long the battery can be powered. The battery we are using can produce 500 mAh for 1 hour. How long a battery last depends on the circuit using it. Let's say that the bike light will use 100 mAh when it is done. With a 500 mAh battery it would then last for 5 hours. Alternatively, calculating the time a battery will last is hard, since it depends on many conditions. Sometimes, your circuits might use more power under intensive operations, and sometimes less power when things connected are not active. The outside temperature can also affect the battery life time, so I would consider any calculations made on battery life time to be rough estimates.

Recharging the battery is possible since this is a lithium-ion battery. In lithium-ion batteries, the ions move between the anode and the cathode. They use a lithium compound as the electrode material instead of metallic lithium used in lithium batteries, which are not rechargeable. You can use any battery really as long as the output voltage is in the range of what the FLORA board can handle and it has a JST connector. The JST connecter is the name of the connector on the opposite side of the mini USB connector.

This is a standard form factor connector, and this type of connector is found on many electronic components. In order to power the FLORA board from an external power source, you need to switch the small switch found on the top of the board over to BAT. The switch also acts as an on off switch while powering your FLORA board from an external power source.

But before you attach your battery, we need to upload the following code to the board. When connecting your FLORA board to the USB cable, the FLORA board will be powered from your computer:

```
#include <Wire.h>
#include <SPI.h>
#include <Adafruit_LSM9DS0.h>
#include <Adafruit_Sensor.h>
#include <Adafruit_TSL2561_U.h>


// i2c
Adafruit_LSM9DS0 lsm = Adafruit_LSM9DS0();
//Give the accelerometer a name
Adafruit_TSL2561_Unified tsl = Adafruit_TSL2561_Unified(TSL2561_ADDR_
FLOAT, 12345);
//Declare names for all the led pins
int redLed1=9;
int redLed2=10;
int whiteLed1=12;
int whiteLed2=6;


void setup()
{
//Declare all the led pins as outputs
  pinMode(redLed1,OUTPUT);
  pinMode(redLed2,OUTPUT);
  pinMode(whiteLed1,OUTPUT);
  pinMode(whiteLed2,OUTPUT);
//Start the serial communication
  Serial.begin(9600);
```

```
//Wait until the communication starts
  while (!Serial);

  // Setup both sensor
  configureSensors();


  if(!tsl.begin())
  {
    /* Cant find the TSL2561, check your connections */
    Serial.print("Cant detect the TSL2561 detected Check your
wiring");
    while(1);
  }
  // Try to initialize the light sensor
  if (!lsm.begin())
  {
    Serial.println("Can not find the sensor, Check your
connections!");


  }

void loop()
{

  /* Get a new sensor event */
  sensors_event_t event;
  tsl.getEvent(&event);

  /* If the lux goes below 70 then its dark and we can start using the
leds  */
  if (event.light>70){
    lsm.read();
    //If the hand is in the turn position light the white leds
    if(lsm.accelData.x<-1000 && lsm.accelData.y<3000){
      digitalWrite(whiteLed1,HIGH);
      delay(500);
      digitalWrite(whiteLed1,LOW);
      digitalWrite(whiteLed2,HIGH);
      delay(500);
      digitalWrite(whiteLed2,LOW);
    }
    //If the hand is in the turn position light the red leds
    if(lsm.accelData.y>1000 && lsm.accelData.x<0){
      digitalWrite(redLed1,HIGH);
```

```
        delay(500);
        digitalWrite(redLed1,LOW);
        digitalWrite(redLed2,HIGH);
        delay(500);
        digitalWrite(redLed2,LOW);

    }
  }
  delay(500);
}
//Configure both the sensors
void configureSensors()
{
  tsl.enableAutoRange(true);
  tsl.setIntegrationTime(TSL2561_INTEGRATIONTIME_402MS);
  lsm.setupAccel(lsm.LSM9DS0_ACCELRANGE_2G);
}
```

In the preceding sketch, we first checked the light conditions. I found that the lux levels were around 70 when the sun started to set, so I used this as my condition for activating the bike glove. The light condition might be different where you are located, so remember to make some tests outside before you choose you own condition. If its dark outside, we then carry on and read the accelerometer to detect which position the hand is in. If the hand is in the turning position, we activate the white LED pattern, and if it is in the stop position, we activate the red light pattern.

Figure 3.7: Everything in place

Once your code is done, it is time to put everything in place. In *Figure 3.7*, you can see how we added a round piece of MDF as a lid. **MDF** stands for **medium density fiberboard**, which is a wooden fiber material that is fairly cheap and easy to cut. I made mine a little bit bigger than the bottom of the cap so that I could press it in for a snug fit. The last thing you need to do is to add Velcro to the bottom of the MDF and stich the opposite piece of Velcro to any pair of gloves, and you are good to go. In order for this turn light system to work, you would have to duplicate the project, so you can fit it to a second glove. But once you have made one, the second should be much easier. In *Figure 3.8* you can see the finished project and how the lighting animation works.

Once you have completed the steps in this chapter, you are free to do any modifications to the project you wish. You could start experimenting with different light patterns, adding more gestures, or experimenting with other materials for the casing. The choice is yours.

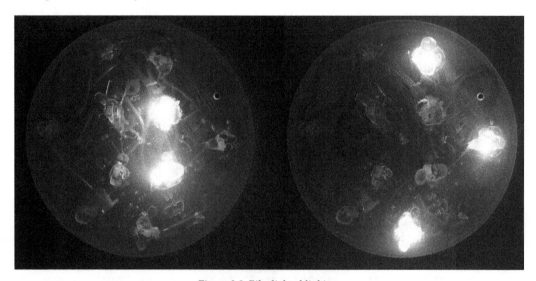

Figure 3.8: Bike lights blinking

Summary

In this chapter, we experimented a bit with the TSL2561 sensor and the accelerometer sensors. The TSL2561 sensor may seem a bit over the top for this project, since you can make the bike gloves using an LDR sensor such as the one in *Chapter 2, Working with Sensors*. Besides being a more accurate light sensor, the TSL2561 sensor has one advantage in regards to this project. Since the pin layout is limited on the FLORA board, we save an analog pin. Using a digital sensor we can communicate with over I2C. The same protocol is what we used to communicate with the accelerometer. Using this sensor, we have also introduced the concept of gesture tracking. This is one possible use of the accelerometer. As your programming knowledge progresses, this sensor might offer new possibilities. We have also learned a bit about how we can finalize projects and how to power them with external power sources. In the following chapters, we will keep on using batteries to power our project and dig deeper into coding.

4
LED Glasses

In this chapter, we will look at how to create a pair of LED glasses. In essence, these glasses are an LED matrix. A matrix is an arrangement of LEDs in columns and rows, where we take advantage of the polarity of the LEDs so we can control 30 LEDs separately using only 15 pins on the Arduino board. The LED matrix has been around for many years and is still used today in some screen technologies. For example, digital bus signs are usually made by implementing an LED matrix, where each individual LED acts like a pixel.

In this chapter you will also learn a little bit about how to create your own glasses and soldering techniques. We will also have a look at some more advanced programming, where we will be implementing some simple animation for you to build upon.

The materials needed are:

- 30 LEDs (5 mm) in any color
- A trinket board or an Arduino micro board
- An FTDI to USB converter
- Wires
- A soldering iron
- A 3-5V lithium battery (the smaller the better)
- A JST female connector
- 3 mm MDF
- 220Ω resistors

Making the glasses

You can make your glasses out of anything you like and it might even be possible to modify a pair of existing sunglasses if they are big enough. My good friend Roger Persson was nice enough to design a pair of glasses for this book. In *Figure 3.1* you will find the design of the glasses with measurements. Remember that these measurements might need to be modified so that the glasses fit your particular head size.

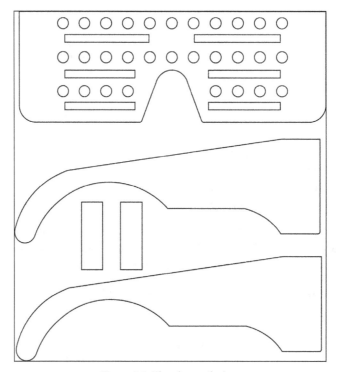

Figure 3.1: The glasses design

The design of the glasses is very simple yet fits the purpose. The design consists of three main pieces, the front and two side frames. The two smaller pieces are used for added support when connecting the frames to the front panel. The frames can be simply glued to the corners of the front panel, while the two small pieces are glued to the inside corners on each end, as shown in *Figure 3.2*. The design template can be used to cut the pieces out from any material you like, but I recommend a sturdy material of at least 3 mm thickness. Plastic materials would also work for this project, or you could even make the frames out of cardboard, which can easily be cut by hand.

The glasses presented in this chapter were cut from 3 mm MDF using a laser cutter. I don't expect most readers to have their own laser cutter, but many bigger cities today have something called hacker spaces or Fab Labs, and some even have them in their city's libraries. The concept of Fab Labs was developed by the Massachusetts Institute of Technology, building on the idea of hacker spaces, which are locations that offer access to different fabrication tools and equipment where different people come to work on projects. Hacker spaces are usually a place where people interested in electronics and programming meet to share ideas and work on projects. Fab Labs are more oriented toward fabrication, not just digital entity, and are open to the public. If you haven't checked already, I suggest you investigate whether there is a hacker space or Fab Labs close by since you now have the perfect excuse to head over for a visit.

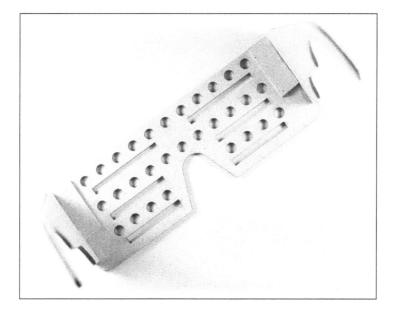

Figure 3.2: The laser-cut pieces

If you choose to modify the design, remember to keep the size of the holes at 5 mm since this is the size of the LEDs we will be using. You can swap these LEDs for the smaller 3 mm ones if you like, but I would not recommend LEDs bigger than 5 mm since these might complicate the design.

Entering the matrix

An LED matrix is also known as a **diode matrix**, referring to the LED's one-directional polarity. An LED matrix is a two-dimensional grid with an LED connected to each intersection where a row crosses a column. The columns and rows are isolated from one another in order for the matrix to work. In electronics this is also known as **multiplexing**.

Figure 3.3 illustrates the entire schematic of all the connections. To the right, you will find the matrix layout. All the negative sides of the LEDs are connected in rows and all the positive sides of the LEDs are connected in columns. When power is applied to one of the columns, and a ground connection is opened up on the negative rows, only one LED will light up. As you might have noticed, we have connected part of the matrix to the analog pins. Since there are not enough digital pins, we will use some of the analog pins instead. The analog pins can be operated as digital pins and their numbering continues on from pin **13**. In other words, analog pin **A1** is the same as digital pin **14**, and so on.

As I said before, the current can only pass through an LED in one direction, a fact we are using to our benefit while creating the LED matrix, giving us the possibility of controlling many LEDs with fewer pins. If we did not connect everything in a matrix, we would need 30 pins in order to control all the LEDs separately. The downside of using a matrix configuration is that we can only control one LED at a time.

However, we will take advantage of another phenomenon called **POV (persistence of vision)**. POV is a term used to describe an optical illusion where multiple images blend into one image in the mind of the beholder. The brain can only interpret about 25 discreet images per second; any more than that and the images start to blend together.

The following *Figure 3.3* illustrates the entire schematic of all the connections:

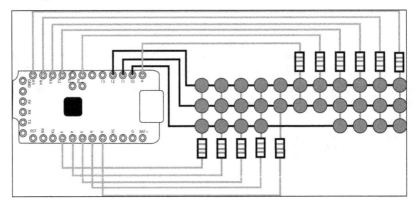

Figure 3.3: The matrix schematic

Arduino is fast, even faster than the human eye, so we will use this speed to our advantage in order to give the impression of lighting up many LEDs at the same time. As I said, we can't technically light up more than one LED at once in the matrix but we can switch between them so fast that the human eye will perceive it as more than one LED being on. But before we get to this part, we need to connect everything, and this means it is time to turn on the soldering iron.

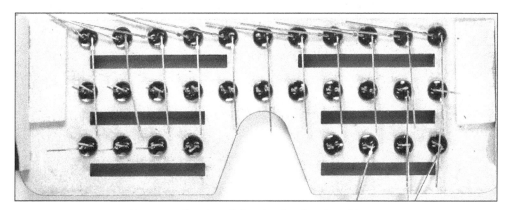

Figure 3.4: Showing all the LEDs lined up for soldering

Before we start soldering, we need to place the LEDs in the right order. A good idea is to check that all your LEDs work before soldering them. By connecting the LEDs one by one to a 3.3V coin cell battery, or using the power that goes from your Arduino to a breadboard, you can easily do this. If you are using a breadboard, don't forget to add a 220Ω resistor.

If you cut 5 mm holes, the LEDs should fit nicely. If they are a bit loose don't worry, as once they are soldered together everything will be held in place. To create the matrix, we need to solder the LEDs into rows and columns. In *Figure 3.4* you can see how I have prepared the LEDs for soldering by bending the legs of the LEDs into the desired rows and columns. All the negative legs (the shorter ones) will be soldered in horizontal lines, and then the positive legs (the longer ones) will be soldered in vertical lines. Make sure that you bend the positive lines over the negative lines so they do not come into contact with one another. If they do, the matrix will not work as it is supposed to. If you want, you can cover up your lines using some tape needed. This is done by placing a small piece of tape in between the legs of the LEDs so they do not touch one another.

Once you are done, we can move on and add the wiring and the resistors that will connect to our Arduino board. *Figure 3.5* shows a close-up of the resistors connected straight to the positive column in the glasses, and the wires connected to the other side of the resistors. The idea is to place the Arduino board on the inside of either the left or right frame. Before you cut your wires, measure the distance between the row and the location of the Arduino board where it will be placed on the inside of the side frame. Make sure you add some extra length before you cut them because it is better to have wires that are too long than too short.

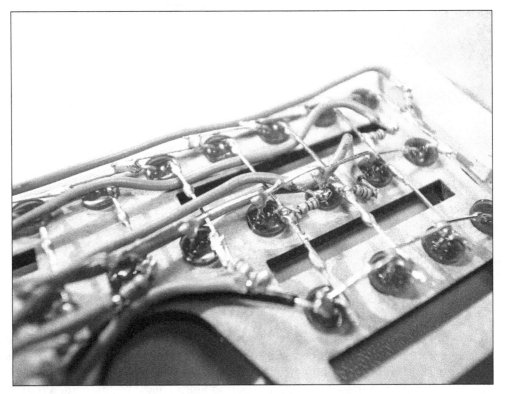

Figure 3.5: A close-up of the matrix

Now is also a good time to put the JST female connector in place as shown in *Figure 3.6*. JST connectors are fairly standard connector for batteries, and in this project we will be using a very small 3.7V battery with a male JST connector. You can place the JST connector anywhere you like, but I found a good spot where the front panel meets the frame just under the supporting piece of MDF. Make sure you leave enough space on the back side of the connector to fit the power cable that connects to the Arduino board. To keep the JST connector in place, use some glue:

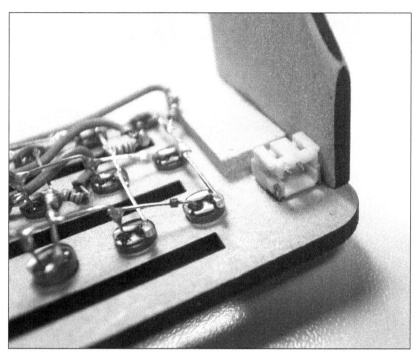

Figure 3.6: The JST female connector in place

When you have added the wires to all the positive columns, you can add three cables for the negative rows, again ensuring you make them long enough to reach the Arduino board. You don't need resistors on these lines since these will act as our **GND** channels.

Once you have all the LEDs, resistors, and wires in place, it is time to connect everything to the Arduino board. In this chapter, we are using the Trinket board from Adafruit, but you could also use an Arduino micro board, which is very similar in size. These boards are some of the smallest Arduino boards that offer most of the functionality of a standard Arduino board.

Soldering all the wires in place might be tricky. I started by gluing the board to the inside of the frame and then soldering the wires one by one. I would suggest that you place them where they fit best. You can always switch the layout in the code later on. In *Figure 3.7* you will see what it might look like once all the wires are connected:

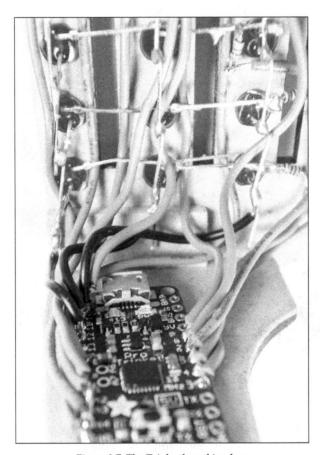

Figure 3.7: The Trinket board in place

Take your time soldering all the wires in place. I admit that even someone with good soldering skills might find this project a bit tricky since it requires some unconventional soldering. I call this type of solder action "soldering" since you usually end up with something that looks like it came from a movie. Usually, you solder components on a flat surface, but with wearable projects like this one you need to be a bit creative when it comes to soldering things together. Eventually you will end up with an inside that is as impressive as the outside.

Next, we will move on to the programming side, and this is where we get to see the glasses in action. For the programming part, we will power the glasses via the USB cable, and once we are done we will add a battery, then you will be ready to head out into the night to impress everyone.

Programming the glasses

In order to make the Trinket board so small, the serial to USB conversion is left out from the design. On a regular Arduino board, this conversion is handled by another Atmega chip, and on older versions this was done by an **FTDI** (**Future Technology Devices International**) chip. The FTDI chips are still around and you can buy these as standalone breakout boards as shown in *Figure 3.8* to the left of the Trinket board:

Figure 3.8: The FTDI serial to USB converter and the Trinket board

Normally, you solder male pins to the end of the Trinket board that connects to the FTDI converter, but in this case we want to keep the Trinket board as flat as possible and we don't want sharp pins on the inside of the glasses that might hurt your eyes. So the trick is to just attach the male pin headers to the FTDI converter and hold it in place while programming the Trinket board. Once in a while there will be glitches in the connection and the upload will fail. This is probably due to the FTDI not connecting properly to the Trinket board. However, this is not a big problem since you can just start the upload over again while making sure the pins have a good connection.

Now let's make a sketch that checks that all of the LEDs light up. In order to do so, we will loop through the LEDs one by one to see that everything works as it is supposed to. The Trinket is programmed as a normal Arduino. Uno board, so make sure you select this type in the board menu Upload the following code and check the LEDs in front of the glasses:

```
/*Collect all the positive columns pins in one array. You need to make
sure that these pins correspond to the direction you have placed the
columns in the glasses*/
int powerPin[]={
   3,4,5,6,8,9,14,16,17,18,19};
/*Collect all the negative row pins in one array. Again make sure they
are added in the same order corresponding to the glasses*/
int gndPins[]={
   10,11,12};

void setup(){
   /*In order for the matrix to work we need to be able to control
our gnd lines in the matrix. The trick is to use all pins as output.
When we turn the gnd pins HIGH we will be able to block the ground
connection*/
   for(int i=0; i<20;i++){
      pinMode(i,OUTPUT);
   }
   //Turn all the gnd pins HIGH in order to keep all LEDs off
   for(int j=0;j<3;j++){
      digitalWrite(gndPins[j],HIGH);
   }
}
void loop(){
   //Run the function
   looper();
}

void looper(){
   /*In this function we run through all the LEDs using two for loops
starting by opening a gnd connection*/
   for(int i=0; i<11;i++){
      digitalWrite(powerPin[i],HIGH);
      //Once a gnd pin is accessible we turn on one of the LEDs
      for(int j=0;j<3;j++){
         digitalWrite(gndPins[j],LOW);
         delay(50);
         digitalWrite(gndPins[j],HIGH);
         delay(50);
```

```
        }
      digitalWrite(powerPin[i],LOW);
    }
  //Loop everything backwards
      for(int i=10; i>=0;i--){
      digitalWrite(powerPin[i],HIGH);
      for(int j=3;j>=0;j--){
        digitalWrite(gndPins[j],LOW);
        delay(50);
        digitalWrite(gndPins[j],HIGH);
        delay(50);
      }
      digitalWrite(powerPin[i],LOW);
    }
  }
```

In this example sketch, we are implementing a trick in order to be able to control the ground connections. If we connected the negative rows of the matrix straight to **GND** we would not be able to control the separate LEDs. The trick is to use normal pins as outputs. When a pin is LOW, it connects to ground, which we can use to light up our LEDs in the matrix. But once it turns to HIGH, we block the connection to the ground. So now we can control each LED individually by turning one of the positive columns HIGH and one of the negatives rows LOW. You will need to make sure that your pin declarations line up with the actual physical layout in your glasses or else looping through them could get very hard. As you can see in the schematic in *Figure 3.3*, the columns are connected after one another to the digital pins.

Making a pattern

In the next code example, we will implement some pattern designs. These patterns are stored in arrays that correspond to the layout of the LEDs in the glasses. We can draw our patterns in code and later loop through the array and activate the LEDs. When the code is formatted as it is in the next sketch, we get a visual repetition of the pattern. A 0 in the array represents a turned off LED in the same position in the matrix and a 1 represent an LED that is turned HIGH:

```
/*Collect all the positive columns pins in one array. You need to make
sure that these pins correspond to the direction you have placed the
columns in the glasses*/
int powerPin[]={
   19,18,17,16,14,9,8,6,5,4,3};
/*Collect all the negative row pins in one array. Again make sure they
are added in the same order corresponding to the glasses*/
int gndPins[]={
```

```
    12,11,10};
//This is a two dimensional array that holds the pattern
int pattern[3][11] = {
   {1,1,1,1,0,0,0,1,1,1,1  },
   {0,1,1,0,0,0,0,1,0,0,1  },
   {0,1,1,0,0,0,0,1,1,1,1  }
   ,
};
//Variable to store the refresh rate on the led display
int refreshRate=200;

void setup(){
  //Declare all pins as outputs
  Serial.begin(9600);
  for(int i=0; i<20;i++){
    pinMode(i,OUTPUT);
  }
  //turn all the gnd ports High to keep them blocked
  for(int j=0;j<3;j++){
    digitalWrite(gndPins[j],HIGH);
  }
}
void loop(){
  //Run the pattern function
  displayPattern();

}
/*Function that runs through all the positions in the pattern array*/
void displayPattern()
{
  for (byte x=0; x<3; x++) {
    for (byte y=0; y<11; y++) {
      int data =pattern[x][y];
    //If the data stored in the array is 1 turn on the led
      if (data==1) {
        digitalWrite(powerPin[y],HIGH);
        digitalWrite(gndPins[x],LOW);
        delayMicroseconds(refreshRate);
        digitalWrite(powerPin[y],LOW);
        digitalWrite(gndPins[x],HIGH);
      }
      //If it is something else turn the led off
      else {
```

```
        digitalWrite(powerPin[y],LOW);
        digitalWrite(gndPins[x],HIGH);
      }
    }


  }
}
```

This sketch implements a two-dimensional array, which is the same as placing an array into an array. As you can see, we have three arrays, and inside each of those arrays we have 11 positions in the first two and eight in the last one, which corresponds to the layout of the matrix. Using the two-dimensional array, we can now fetch the positions of the LEDs similar to x and y coordinates, which is much easier than storing everything in a normal array. If the values are stored in a normal array, we would need to define where each row ends on our own. This could be done using `if` sentences to check where the row begins, but using a two-dimensional array makes things much easier and makes for better-looking code.

Then the arrays run through the display pattern function, which loops through all the positions in the array. Every time it finds a 1 in the array it turns on the LED corresponding to the position in the actual glasses. It only turns it on for a brief time based on the refresh rate before it turns it off, since we can only have one LED on at a time in the LED matrix. Again, this is where we use the POV phenomenon, looping through all the LEDs very fast so that when we look at the glasses it looks like multiple LEDs are on, though in fact there is only one LED on at a time.

In order to get a better understanding of the code, I would suggest you modify the pattern array by changing which LEDs light up. If you look closely at the array, you might make out that I have tried to spell my initials TO with 1s in the code, which corresponds to LEDs turned on. Try switching the letters for your own initials and upload the code to your glasses.

Finishing the glasses Knight Rider style

For the last code example, we will have a look at how to create an animation. An animation is a simulation of motion, and in one sense the first code example in this chapter is a form of animation. We will build on the same principle in this section. Once you get the hang of the basic concepts, you can start building your own animations, combining the knowledge from the pattern example with the knowledge in this sketch.

In my beginner Arduino classes, the Knight Rider example is a classic. This example is inspired by the 80s hit show *Knight Rider* with David Hasselhoff. To be more precise, the example is inspired by the robotic car featured in the show, which is called Kit. In the front of Kit there is a small LED display that shows a bouncing light effect. This is the effect we will recreate on the front of the glasses.

The code example is fairly simple and does the same thing as the test sketch, but instead of lighting up only one LED at a time we will light up an entire column. We will move the light column from left to right and once we hit the end of the glasses, we will move the column back again:

```
int powerPin[]={
   19,18,17,16,14,9,8,6,5,4,3};

int gndPins[]={
   12,11,10};

int refreshRate=200;

void setup(){
  for(int i=0;  i<20;i++){
    pinMode(i,OUTPUT);
  }
  //
  for(int j=0;j<3;j++){
    digitalWrite(gndPins[j],HIGH);
  }
}
void loop(){
  nightrider();

}
//
void nightrider()
{
  /*Instead of starting to loop through the columns we loop through
the row*/
  for(int i=0;  i<11;  i++){
/*Then we loop through the column
    for(int j=0;  j<3;  j++){
/*In order to perceive that the column is lit we need to loop it a few
times*/
      for(int k=0;  k<50;  k++){
/*Then we light the column
```

```
            digitalWrite(powerPin[i],HIGH);
            digitalWrite(gndPins[j],LOW);
            delayMicroseconds(refreshRate);
            digitalWrite(powerPin[i],LOW);
            digitalWrite(gndPins[j],HIGH);
        }
      }
    }
    /*Once we have reached the end of the glasses we do the same thing
    backward*/
    for(int i=11; i>0; i--){
      for(int j=3; j>0; j--){
        for(int k=0; k<50; k++){
            digitalWrite(powerPin[i],HIGH);
            digitalWrite(gndPins[j],LOW);
            delayMicroseconds(refreshRate);
            digitalWrite(powerPin[i],LOW);
            digitalWrite(gndPins[j],HIGH);
        }
      }
    }
}
```

Once your code is done, upload it to the board and enjoy. In order to show off your new glasses, we need to finish up the circuit by attaching the battery to the glasses so you can walk around with them (since being forced to be connected to a USB ports at all times might not look as cool). Any battery between 3-5V will work for this project, and the more amperes the battery has, the longer the glasses will stay on. However, large ampere batteries are also bigger. The smallest lithium battery I have found is the 3.7V 150mAh battery, which I recommend. There are smaller ones with less amperes but this type will still fit your glasses and give you enough power to keep the glasses on long enough for you to impress a few people.

Before you connect the battery, you need to solder wires from the JST connector to the board. The pins for external power on the Trinket board are marked **BAT+** for the positive connection. Next to this pin there is a pin marked with only a **G** for the ground connection. This connection is connected to the negative side of the JST connector. In order to figure out which pin is which on the JST connector, check with the battery cables. Usually, these cables are red and black; the red one is the positive and the black one is the ground. The design of the glasses should leave enough space on the inside of the frames to place a small battery. *Figure 3.9* shows what it might look like. I recommend gluing the battery in place using hot glue, but be careful not to heat the battery for too long or it might explode. This might sound scary, but you would need to heat it for some time before it explodes. However, I always recommend caution when combining heat and batteries.

As you might have noticed by now, visibility is limited in these glasses and it has to be fairly dark for the light to be visible, so be careful when using them. You can still look cool while keeping safe.

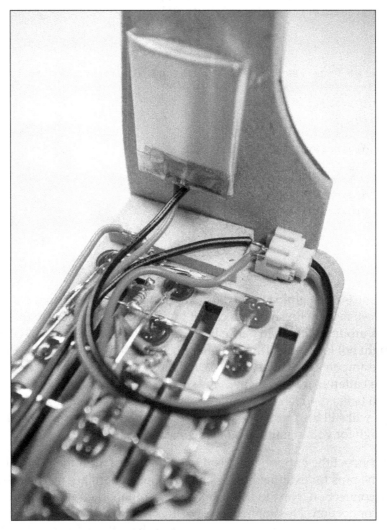

Figure 3.9: Battery connection

In *Figure 3.10* you will find my co-worker Johannes at the university showing off the finished glasses. Unfortunately, pictures do not show the full effect of animating the LEDs, but hopefully you will be close to finishing your pair by the time you read this.

Figure 3.10: Johannes Nilsson showing off the LED glasses

Summary

In this chapter, you learned how to create an LED matrix in a pair of custom-made LED glasses. The principles behind the LED matrix are the same as in any matrix that you might use in another project. On the code side, we had a look at some pattern designs and how to generate animations. This chapter introduces the basic concepts, but as you progress you can build upon this knowledge and develop your own patterns and animations.

In this chapter, we also had a look at what the custom casings looks like when cut using a laser cutter. If you do not have access to a laser cutter don't worry, as you can still achieve the same results using cutting tools and materials such as plastic, wood, or cardboard if you like. The only difference is that it might take a bit longer to cut. The material for the frame also gives you a lot of personalization options, so I recommend you test a few materials before you decide.

The remaining projects will also implement laser-cut designs, but remember that the casings for all the projects in this book are just examples of how it can be done. I encourage you to develop your own designs since it is more fun to personalize your creations, and let the designs in this book act as inspiration.

5
Where in the World Am I?

I hope you enjoyed the projects so far, but now its time to get into some serious prototyping. Does the idea of connecting your project to satellites orbiting the earth tingle your creative senses? It certainty gets me going, and this is just what we will be doing in this chapter. The goal is to make your own a watch, which besides telling time, it will also give you your exact GPS position, with some added information about how fast you are traveling and your altitude above sea level. **GPS** stands for **Global Positioning System**. The development of GPS started in the 1970s, but was not completed until 1992. Today there are 27 satellites orbiting the earth. These satellites give anyone with a GPS receiver the possibility to access their own position, in coordinates of longitude and latitude. Today, you can find GPS in a range of devices for cars, airplanes, boats, and soon in your own watch.

In order to be able to actually see the information, we will introduce the use of LED screens. LED screens come in all shapes and sizes, and in this chapter we will be using one of the smallest one. We will learn how to receive data from a GPS receiver and to display it on the screen. We will also take a look at how to create your own watch casing. The watch we will be making will be slightly bigger then a normal watch, but since it is a GPS watch you made by yourself, your really want it to stand out anyway. In *Figure 5.1*, you will find the modules needed for this chapter:

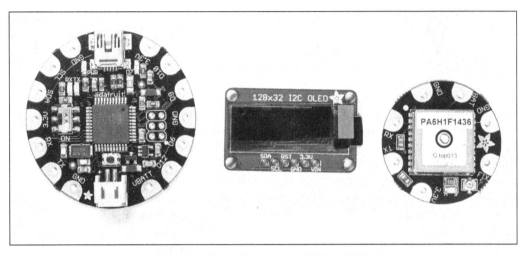

Figure 5.1: The FLORA board, OLED and GPS

We will stick with the FLORA board for this project, mainly because of the round form factor, which will be used to base the design of the watch on. To the right of the FLORA board we have an OLED screen that displays 128 x 32 pixels. We also have a GPS breakout board based on the MTK3339 chipset. In addition to these components, you will need the following materials:

- Soldering iron
- Wires
- Cable cutters
- 3 mm MDF or 3 mm Acrylic plastic (shown in *Figure 5.6*)
- 3.3V 150mA lithium battery
- Fabric straps
- Velcro

Hocking up the OLED screen

The screen we will be using for this project is a 128 x 32 pixel monochrome OLED screen based on a SSD1306 driver. **OLED** stands for **organic light-emitting diode**. The technology is based around a type of coal, which is the organic part of the screen. The benefits with OLED technology is that these screens do not require back light in order to display information. This makes them thinner than normal LCD screens, which are the most common small displays. OLED screens can also achieve a much higher contrast than regular LED screens. The screen we will be using is a monochrome screen, which means that it's only black and white, but there are color OLED displays available too.

To help ease the programming part of the project, we will be using a great library written by Limor Fried for the SSD1306 driver. Limor Fried is also the founder of Adafruit, the company that has designed many of the boards and components used in this book. You can find the library at `https://github.com/adafruit/Adafruit_SSD1306` by pressing the download zip button, then unzip ping the folder and installing it to you library folder. If you have forgotten where your library folder is placed, you can find the path in the Arduino IDE setting by navigating to **Preferences** | **Options**. Don't forget to rename the folder `adafruit_sdd1306`.

Limor has also developed another library that provides different graphical components such as lines, circles, points, and so on. This is a great resource and will be using these graphical components later on. You can find the library at `https://github.com/adafruit/Adafruit-GFX-Library` and install it using the same method as used for the previous library.

Before we get started with the code, let's take a look at how we hook up the screen to the FLORA board. The connections are fairly easy since we will be communicating with the screen over I2C, which is the supported communication protocol for this device. If you need to refresh you memory on I2C, take a look at *Chapter 3, Bike Gloves*.

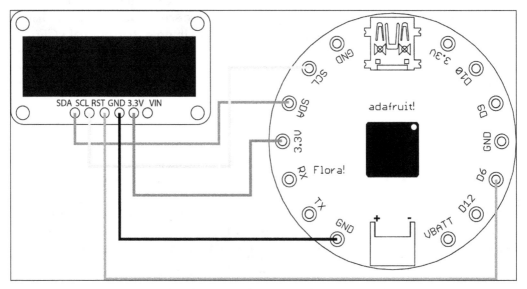

Figure 5.2: Connecting the OLED to the FLORA board

In *Figure 5.2*, you will find the necessary connections. The **SDA** and **SCL** pins connect to one and other, as do the **3.3 V** and **GND** pins. The screen has a reset pin, which we will connect to pin number 6 on the FLORA board. For testing the screen, I recommend soldering wires to your screen and hooking it up to the FLORA board using alligator clips. If you make the wires about 10 cm long, you can then use them for soldering everything to the FLORA board once you are done testing.

Now, let's try something simple to check if you screen is working properly. In the next sketch example, we will just display random pixels in order to test that everything is working:

```
//Add the nessesary libraries
#include <SPI.h>
#include <Wire.h>
#include <Adafruit_GFX.h>
#include <Adafruit_SSD1306.h>
//Set the reset pin on the FLORA board
#define OLED_RESET 6
Adafruit_SSD1306 display(OLED_RESET);
```

```
void setup()   {
  Serial.begin(9600);
  //initialize the screen with the I2C addr 0x3C
  display.begin(SSD1306_SWITCHCAPVCC, 0x3C);

}


void loop() {
   //Clear the screen buffer
    display.clearDisplay();
   //Draw a random pixel on the screen and make it white
    display.drawPixel(random(0,128), random(0,32), WHITE);
   //Send the information to the screen
    display.display();
   //Wait for a bit
    delay(50);
}
```

The preceding sketch will display a pixel in white for 100 milliseconds before it shows a new one. Keep in mind that any graphics or information sent to the screen is stored on the screen until you clear it. This means that if you display something on the screen and then turn it off and on again, the same image will appear. This is why we first clear the screen of any information at the start of the sketch. Then we can add all the information we want, and this is not displayed on the screen until we send the display.display() command.

Upload the sketch to you FLORA board, and if you see white pixels jumping around, everything is working fine and we can try the following example. In the next sketch, we will make a simple watch in order to see how we display information on the screen:

```
#include <SPI.h>
#include <Wire.h>
#include <Adafruit_GFX.h>
#include <Adafruit_SSD1306.h>
//Make sure to set the resetpin to the one you are using
#define OLED_RESET 6
Adafruit_SSD1306 display(OLED_RESET);

//Variables for the clock
int s=0;
int m=0;
int h=0;
```

```
void setup()    {
  Serial.begin(9600);

  //initialize with the I2C addr 0x3C (for the 128x32)
  display.begin(SSD1306_SWITCHCAPVCC, 0x3C);

}


void loop() {
  //Start the counter for the stop watch
  s++;
  //Clear the screen
  display.clearDisplay();
  //Set the text size (1 is the smallest size)
  display.setTextSize(3);
  //Set the color of the text
  display.setTextColor(WHITE);
  //Set the starting position for the text in x and y coordinates
  display.setCursor(0,0);
  //Print the information to the screen
  display.print(h);
  display.print(":");
  display.print(m);
  display.print(":");
  display.print(s);
  //If the counter becomes bigger then 59 eg 1 minute make it 0 again
  and increase the minutes by one
  if(s>59){
    s=0;
    m++;
  }
  //If the minute counter becomes bigger then 59 eg 1 hour make it 0
  again and increase the hour by one
  if(m>59){
    m=0;
    h++;
  }
  //If the minute counter becomes bigger then 59 eg 1 hour make it 0
  again and increase the houre by one
  if(h>23){
    h=0;
  }
```

```
    //Send everything to the screen
    display.display();
    //Wait for a second before we update everything
    delay(1000);
}
```

As you can see it the preceding sketch, we can set the size of the text and where we want to display it on the screen. If you change the size to 1, you can fit up to four lines of text on the screen you are using, which is prefect since there is a lot of information available from the GPS receiver.

Getting the position

Now it's time to take a look at the GPS receiver. The GPS we will be using is based on the MTK3339 chipset, which is very easy to use but still very powerful. Adafruit manufactures the board shown in *Figure 5.1*, and this module was built with wearables in mind. This GPS can track up to 22 satellites and has very low power consumptions, which makes it ideal for battery-powered projects. The GPS is incredibly easy to read, you just need to connect over the serial port and you are good to go. In *Figure 5.3*, you will find the necessary connections that need to be made. I recommend connecting everything using alligator clips while you are trying out the GPS receiver:

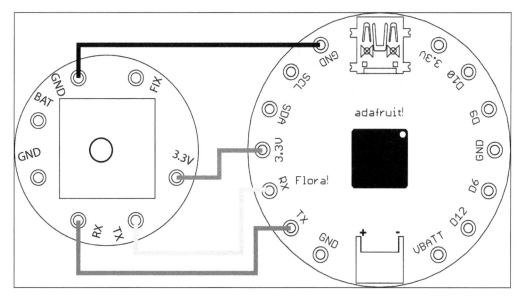

Figure 5.3: Connecting the GPS to the FLORA board

Don't forget to connect the serial pins in the right order. The **TX** pin of the GPS needs to connect to the **RX** pin on the FLORA board and vice versa, since we want to transmit from the transmitting pin to the receiving pin. The GPS module will receive the position data. In order to display it on the OLED screen, we need to pass it on to the FLORA board for processing. Once you are done, you can upload the following code. Once uploaded, you can open your serial monitor and the information should start flowing:

```
void setup() {
  //Setup the usb serial
  Serial.begin(9600);
  //Set up the FLORA serial
  Serial1.begin(9600);
}


void loop() {
  //If there is data coming in on the FLORA serial port
  if (Serial1.available()) {
    //Store it
    char c = Serial1.read();
    //Print it back over the usb serial
    Serial.write(c);
  }
}
```

As you can see in the setup, we declared two serial ports. This is because the FLORA board has two separate serial ports. On the standard Arduino board, the **RX** and **TX** pins are connected to the same serial port as the USB, but on the FLORA board, they are two separate ports. In order to read the information sent from the GPS, we first have to pick up on the FLORA board and then send it back to the serial monitor.

The down side of this GPS is that the receiver sends information continually over the serial port, so that at a glance it can be hard to tell what information is relevant. The information is sent in four different packages, each starting with a dollar sign. Parsing this data flow is tricky business and will take too long to explain in this chapter, but as usual, the components from Adafruit come with a great library for sorting out the data, and this one is also written by Limor Fried. You can find the library at `https://github.com/adafruit/Adafruit-GPS-Library`, just download it and install as you did with the ones before.

When a GPS receiver connects to a satellite and gets the information about its position, this is also known as getting a fix. There are a lot of things that can interfere with the GPS receiver, such as static electricity or other objects, and getting a fix indoors can be hard sometimes. In some cases, it can even take up to 45 minutes before a receiver gets a fix on a satellite, so some patience may be required. A good tip is to move around until you find a spot that allows you to get a fix on you position, and use this spot for future debugging. The ideal place to get a fix is outdoors, with clear visibility from the top of the GPS and the sky. Upload the following sketch and take a look in you monitor, if you can get a fix. The following example sketch will print all the information available from the GPS receiver:

```
//Add the libraries
#include <Adafruit_GPS.h>
#include <SoftwareSerial.h>
//Connect the GPS to the FLORA serial port
Adafruit_GPS GPS(&Serial1);

void setup()
{
  /* connect at 115200 serial to USB at a high speed so information
does not get dropped*/
  Serial.begin(115200);
  //Send a test message
  Serial.println("Testing GPS");

  // 9600 NMEA is the default baud rate for MTK3339
  GPS.begin(9600);
  // Set the update rate of the GPS
  GPS.sendCommand(PMTK_SET_NMEA_UPDATE_1HZ);
  delay(1000);
}
//Make a timestamp for our timer
uint32_t timer = millis();
void loop()
{
  // read data from the GPS
  char c = GPS.read();


  // if there is data we parse the information
  if (GPS.newNMEAreceived()) {
    //
    if (!GPS.parse(GPS.lastNMEA()))
      return;
```

```
  }
  // If the timer goes wrong reset it
  if (timer > millis()) timer = millis();

  //if the timer is bigger than 2 sec print the data
  if (millis() - timer > 2000) {
  // reset the timer
    timer = millis();
    Serial.print("\nTime: ");
    Serial.print(GPS.hour, DEC);
    Serial.print(':');
    Serial.print(GPS.minute, DEC);
    Serial.print(':');
    Serial.print(GPS.seconds, DEC);
    Serial.print('.');
    Serial.println(GPS.milliseconds);
    Serial.print("Date: ");
    Serial.print(GPS.day, DEC);
    Serial.print('/');
    Serial.print(GPS.month, DEC);
    Serial.print("/20");
    Serial.println(GPS.year, DEC);
    Serial.print("Fix: ");
    Serial.print((int)GPS.fix);
    Serial.print(" quality: ");
    Serial.println((int)GPS.fixquality);
    //If we get a satellite fix
    if (GPS.fix) {
      Serial.print("Location: ");
      Serial.print(GPS.latitude, 4);
      Serial.print(GPS.lat);
      Serial.print(", ");
      Serial.print(GPS.longitude, 4);
      Serial.println(GPS.lon);
      Serial.print("Speed (knots): ");
      Serial.println(GPS.speed);
      Serial.print("Angle: ");
      Serial.println(GPS.angle);
      Serial.print("Altitude: ");
      Serial.println(GPS.altitude);
      Serial.print("Satellites: ");
      Serial.println((int)GPS.satellites);
    }
  }
}
```

As if making a GPS watch was not cool enough, your watch will be the most accurate wristwatch there is. As you can see in the beginning of the messages that are being printed out in the sketch, we are printing the time and date. Now, you may wonder how the GPS receiver knows what time it is. Well, GPS satellites carry atomic clocks, which are the most accurate clocks there are. These clocks are synced everyday with ground clocks on earth and the other satellites, and any drift from true time is corrected. The time is displayed in **Coordinated Universal Time (UTC)**, which is the standard for telling time. This timeline is based in Greenwich, in London, and it is from this timeline that time zones are based. If you want your local time, you need to add or subtract hours to or from the Greenwich time.

If the GPS gets a fix, it will also print out data regarding your longitude and latitude, as well as your altitude, the number of satellites we can find, and your speed measured in knots per second. Now we have a screen and the information to display it, so it is time to put everything together.

Making the clock

As we progress through the book, some of the projects will become smaller and smaller in size, and this is no exception. Before we get started with putting everything together, I would just like to make a friendly note on planning the construction of your clock. The next part is a step-by-step guide on how to construct your watch. However, it might be a good idea to read through everything before you get started. Even if this watch is slightly bigger then a normal watch, it can still be tricky to fit all the components in. It should be possible, but if you find it too hard you can always modify the design of the watch and scale it up a bit.

First, we have the schematic of all the connections we need to make. In order to fit everything inside the watch, we will need to solder the connections with wires, as shown in the following *Figure 5.4*:

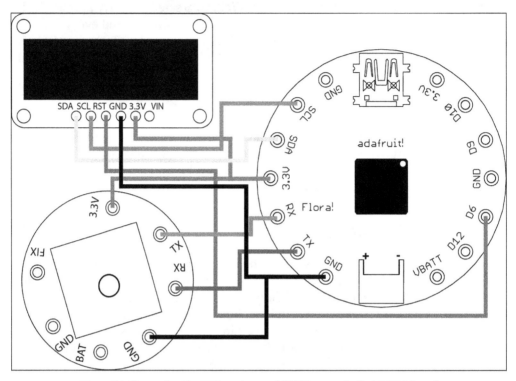

Figure 5.4: Connecting the GPS receiver and OLED screen to the FLORA board

In *Figure 5.4*, you will find the schematic. But before you start soldering everything, you need to make a case for your watch. In *Figure 5.5*, you will find the design of the case with measurements. The design was made for laser cutting using 3 mm MDF or acrylic plastic, as shown in the following *Figure 5.5*:

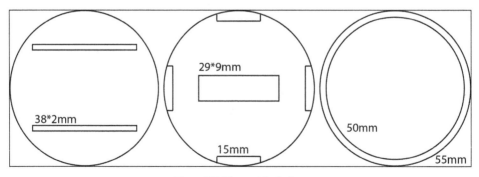

Figure 5.5: The watch design

The idea is to make the casing in layers, which are glued together so that you can cut the design in any material or just use it as inspiration for your own case design. On the left in *Figure 5.5*, you will find the base plate with two rectangular holes. These holes are for running the fabric straps through the casing, which in my case, has a width of 38 mm. A thinner strap will also work as long as you make sure to modify your design.

In the middle of *Figure 5.5*, you will find the top plate. The hole in the middle of this piece is for the OLED screen. The hole only measures the size of the visible parts of the screen, and not the entire module. The last piece on the right is one of the rings that make the actual casing. These are stacked upon one and other and glued together. Depending on your material thinness, you will need a few of them to cover the electronics. All the pieces are 55 mm in diameter, and make sure to keep the four-corner pieces left over from the top piece, since we will use them to hold the top in place, as shown in the following *Figure 5.6*:

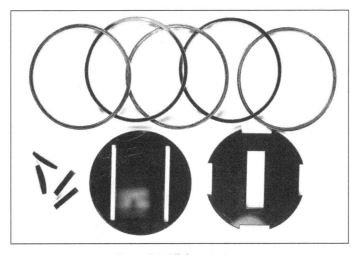

Figure 5.6: All the cut pieces

Once you have cut all the pieces, you should end up with something that looks like *Figure 5.6*. Then it's time to start gluing everything together. I recommend using super glue if you are using acrylic plastic, or normal wood glue if you are using MDF or any other wooden material. Start by gluing one of the rings to the base plate and as soon as it dry, add another one and so on until you think you have enough space on the inside, to fit the electronics. You might need to add another one once we have soldered all the electronics together.

For now, just make a rough estimate based on stacking all the electronics together inside the casing. Don't forget to measure the battery as well:

Figure 5.7: the case glued together

Once done, you should end up with something that looks like *Figure 5.7*, and you can go on and fit the strap into the case. In *Figure 5.6*, the corner pieces left over from the top cut have been glued in place. I recommend waiting before gluing these pieces in place until you are sure that everything will fit inside, just in case you need to add an extra ring.

Figure 5.8: Attaching the OLED display

Soon it will be time to turn on the soldering iron, but before we do so, you need to glue the OLED screen into place, as shown in *Figure 5.8*. For this, I recommend using hot glue, since this will not damage your screen and it gives you some room for errors. Make sure the display is visible from the other side before you glue it into place.

Now its time to solder everything together. Take a good look at the schematic in *Figure 5.4* before you get started in order to get a sense of how to solder everything together. I first placed the GPS receiver on top of the FLORA board and then soldered all the connections between them before attaching the OLED screen. *Figure 5.9* shows how to place everything so it stacks up nicely:

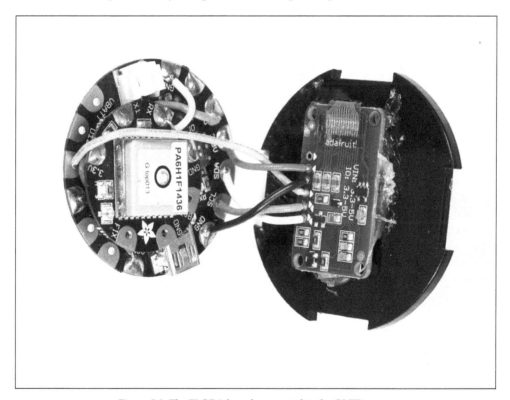

Figure 5.9: The FLORA board connected to the OLED screen

The idea is to have the battery located under the FLORA board once everything is placed inside the case. Once you are sure everything will fit inside the case, you can go ahead and glue the last four corner pieces into place. If the measurements are correct, it will almost give you a press fit so that the lid will be held in place with out any glue. You want to avoid gluing the to the case, since you will need to recharge the battery once in a while and you might want to explore some more creative code for the display.

The final sketch

By now you should have your case ready and all your components soldered, so it is time to add the final sketch to the FLORA board. The following sketch is basically a combination of the GPS and OLED sketches we tried out before. Once you are done, you can upload it to the board. If everything is correct, the information should be displayed on the screen:

```
#include <Adafruit_GPS.h>
#include <SoftwareSerial.h>
#include <SPI.h>
#include <Wire.h>
#include <Adafruit_GFX.h>
#include <Adafruit_SSD1306.h>
Adafruit_GPS GPS(&Serial1);

#define OLED_RESET 6
Adafruit_SSD1306 display(OLED_RESET);

void setup()
{
Serial.begin(115200);
  display.begin(SSD1306_SWITCHCAPVCC, 0x3C);
  // Clear the buffer.
  display.clearDisplay();
  Serial.println("Testing GPS");

  // 9600 NMEA is the default baud rate for MTK3339
  GPS.begin(9600);

  // Set the update rate of the GPS
  GPS.sendCommand(PMTK_SET_NMEA_UPDATE_1HZ);

  delay(1000);
}

uint32_t timer = millis();
void loop()
{
char c = GPS.read();

if (GPS.newNMEAreceived()) {

    if (!GPS.parse(GPS.lastNMEA()))
```

```
        return;
    }

    if (timer > millis()) timer = millis();


    if (millis() - timer > 1000) {
      timer = millis();
      display.clearDisplay();
      display.setTextSize(1);
      display.setTextColor(WHITE);
      display.setCursor(0,0);
      display.print(GPS.hour, DEC);
      display.print(':');
      if(GPS.minute<10){
        display.print('0');
        display.print(GPS.minute, DEC);
      }
      else{
        display.print(GPS.minute, DEC);
      }
      display.print(':');
      if(GPS.seconds<10){
        display.print('0');
        display.print(GPS.seconds, DEC);
      }
      else{
        display.print(GPS.seconds, DEC);
      }
   display.print(" ");
      display.print(GPS.day, DEC);
      display.print('/');
      display.print(GPS.month, DEC);
      display.print("/20");
      display.println(GPS.year, DEC);
if (GPS.fix) {
display.print("Longitude:");
      display.print(GPS.latitude, 4);
      display.println(GPS.lat);
      display.print("Latitude:");
      display.print(GPS.longitude, 4);
      display.println(GPS.lon);
      display.print("Speed:");
      display.print(GPS.speed);
```

```
        display.print(" Alt:");
        display.println(GPS.altitude);
    }
  display.display();


  }
}
```

Once you have uploaded the code and checked that everything is working, you can attach the battery and place everything inside the casing. This design does not have an on–off switch, but instead uses the battery switch on the FLORA board to turn the clock on and off, which is another reason for not gluing the top to the case.

When everything is in place, you should end up with something that looks like *Figure 5.10*. In the first line of the display, the time and date are shown. The second and third lines show the longitude and latitude. The last line shows the speed and altitude. Unfortunately, I could not get a fix inside my photo studio, so there is no actual position displayed in *Figure 5.10*, but by now you probably have the real thing in front of you.

The last thing to do is to add some Velcro to the straps, or you could also use a normal watchstrap if you have one lying around. However, once you show your watch off, the last thing people will focus on is the strap:

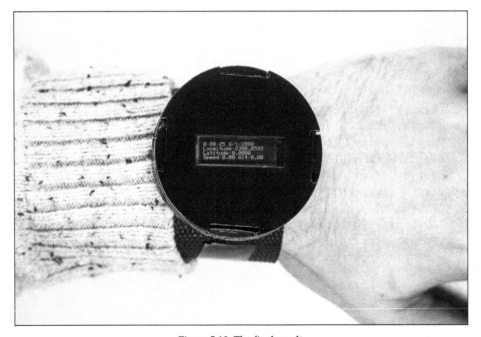

Figure 5.10: The final result

Summary

In this chapter, you learned about how to interact with small screens in the form of an OLED display. We have learned how to display text on the screen and how to program a simple clock. We have also played around with GPS receivers and how to gather information, such as the position of the receiver, speed and altitude from satellites. Then we took a look at how you can design your own watchcase.

There is still tons of fun stuff you can do with displays and if you want to learn more I suggest you check out the example code that comes with the SSD1306 library, which you can find under examples in you Arduino IDE if you have installed the library. For more information about the MTK3339 GPS receiver you can try searching online for the product name and datasheet. Most electronic components are supplied with a datasheet that contains detailed information about it, but may be tricky to read if you are a newcomer.

A good place to start when developing projects in the future is to think of new ways to display the information on the screen. For example, you can make the font bigger and switch between the different types of information that can be displayed, instead of displaying everything at once.

As always, the case design shown in this chapter is just a suggestion, and I urge you to improve upon it in any way you like. Make your own watch case look as good as a high-end watch if you can. But if you ask me which one is better, an expensive gold watch or a watch that connects to a satellite in outer space? The answer is easy.

6
Hands-on with NFC

In this chapter, we will be taking a closer look at **NFC**, which stands for **Near Field Communication**. NFC technology is a means to transfer data wirelessly over short distances. The name comes from the antenna used that usually does not permit communication between devices further than 10 cm. So, what is the benefit of using it? Well, the forced close communication range can be used to create secure and innovative interactions between devices, since you need to be physically present in order to exchange the information.

You can find NFC technology in many modern phones where it is used for a variety of applications, but the most common is probably as a payment system. With NFC on your mobile phone, you can make purchases by simply holding up your phone to a receiver on a cash register. Many public transports around the world also use NFC for paying on buses and trains, but then the NFC is in a plastic card.

Some more modern applications include keyless doors. To be honest, this is not a new invention, but it is only in recent years that they have been commercially available. For example, hotels have been using them for some time. Doors will also be the basis for this chapter's project. Since the theme of this book is wearables, basing a project on doors might seem odd, but don't worry. We will also take a look at how to create your own NFC enable "bling" that can be worn, which will also be used as our keyless keys. To complete the project, you will need the following components found in *Figure 6.1*:

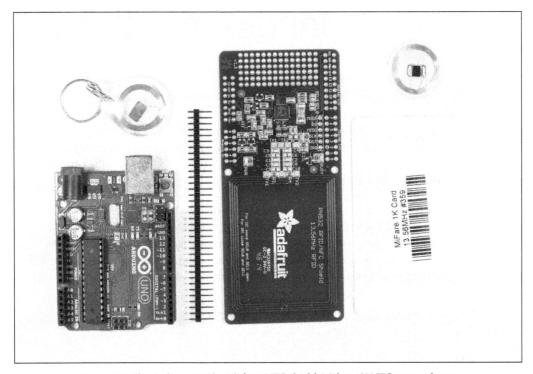

Figure 6.1: The Arduino UNO, Adafruit NFC shield, Mifare 1K NFC tags and cars

In addition to the components shown in *Figure 6.1*, you need the following materials:

- 6 mm MDF board
- Wood or other solid materials for the ring
- USB cable
- Soldering iron
- Wood glue
- Standard servo motor
- Battery or transformer 7–12V
- Male pin headers

Reading a card

Sometimes people also refer to NFC as **RFID**, which is a similar older technology. **RFID** stands for **radio frequency identification** and both NFC and RFID employ radio signals for tagging all sorts of objects. NFC is a newer technology, which includes the same, read and write feature of RFID but has two more functions, which involve card emulation and P2P (peer-to-peer) communication. In this chapter, we will focus on the read and write function. The tags and the card show in *Figure 6.1* have a small memory of 1K where a unique number is stored. The wonderful part of NFC/RFID technology is that the tags and cards do not need any batteries to work. These tags and cards are powered from the signal of the receiver, which gives them enough "juice" to broadcast the information stored in the memory. This is why these tags and cards can't be read if they are further away than 10 cm from the receiver. The electromagnetic field simply can't travel that far. But still, this gives the technology some interesting capabilities in terms of the interactions we can create with them.

Before we get started with the code, we will need to solder the pins to the NFC shield. Any PCB that can be placed on an Arduino board is usually called a **shield**. When it comes to soldering pins to an Arduino shield, there is a common trick you can use. First, start by dividing the pins for the corresponding pins of the shield, which matches the pin layout of the Arduino board. Then place the pins into the Arduino board pin holes and place the shield on top. In this way, everything is kept in place while you are soldering the connection on the top. Once you are done, it should look something like *Figure 6.2*.

Figure 6.2: The pins solder into place on the NFC shield

Once done with the soldering, let's gets started with some code to see if we can read one of the tags. The NFC shield we are using for this project comes with a library by Limor Fried and Kevin Townsend. You can find the library at: `https://github.com/adafruit/Adafruit-PN532`.

Just download and install it as done in previous chapters. Don't forget to rename the folder to `Adafruit_PN532`.

Once the library is in place, use the following code to enable the reading of tags and cards:

```
#include <Wire.h>
#include <SPI.h>
#include <Adafruit_PN532.h>
// Define the I2C pins
#define PN532_IRQ    (2)
// Not connected by default on the NFC Shield
#define PN532_RESET (3)

//Connect the shield I2C connection:
Adafruit_PN532 nfc(PN532_IRQ, PN532_RESET);

void setup(void) {
  //Start serial communication
  Serial.begin(115200);
  Serial.println("Starting the NFC shield");
  nfc.begin();
  //Check the firmware version of the shield
  uint32_t versiondata = nfc.getFirmwareVersion();
  if (! versiondata) {
    Serial.print("Cant find the shield, check you connection");
    while (1); // wait
  }
  // Got ok data, print it out!
  Serial.println("Connected");
  //Print out the firmware version
  Serial.print("Firmware ver. ");
  Serial.print((versiondata>>16) & 0xFF, DEC);
  Serial.print('.'); Serial.println((versiondata>>8) & 0xFF, DEC);
  // Set the number of retry attempts to read from a card
  nfc.setPassiveActivationRetries(0xFF);
  // configure the board to read RFID tags
  nfc.SAMConfig();
  Serial.println("Waiting for a tag");
}
void loop(void) {
  //Decalre a variable to store state
  boolean success;
  //Buffer to store the ID of the card
  byte uid[] = { 0, 0, 0, 0, 0, 0, 0 };
  //Variable to store the length of the ID
  byte uidLength;
```

```
// Waits for and tag and reads the lenght of the ID and the ID
 success = nfc.readPassiveTargetID(PN532_MIFARE_ISO14443A, &uid[0],
&uidLength);
 if (success) {
   //If a card or tag is successfully read print the information
   Serial.println("Tag registered");
   Serial.print("ID Length: ");
   Serial.print(uidLength, DEC);
   Serial.println(" bytes");
   Serial.print("UID Value: ");

   for (uint8_t i=0; i < uidLength; i++)
   {
    Serial.print(uid[i]);
    Serial.print(",");
   }
   Serial.println();
  // Wait 1 second before continuing
  delay(1000);
  }
}
```

Once the code is uploaded to your Arduino board, open up your serial monitor and place a tag or card near the NFC shield. If everything works as it's supposed to, the monitor should receive a message that looks like the one in *Figure 6.3*. As you can see, the UID (unique identification number) stored on my tag is a 4 byte number, which is 39, 246, 64, and 175.

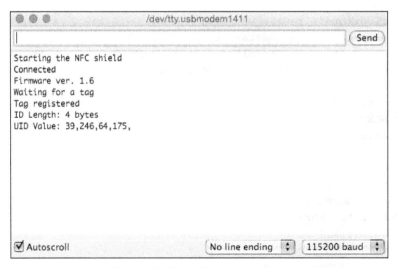

Figure 6.3: The serial monitor output

If you have more then one tag or card, you can check the ID of all of them to see if there are any faulty ones in your mix. Then it is time to pick one of your tags, which we will continue using for the rest of the chapter. I picked the transparent circle tag shown above the NFC card in *Figure 6.1*. The reason for this is that later I will use this tag in the design of my "keyring," but of course, you can choose any tag available to you as long as you can extract the ID number from it.

Write down your number, since we will need it for the next code example where we implement a check into the code that will act as the out lock/unlock mechanism.

Upload the following code to your Arduino board:

```
#include <Wire.h>
#include <SPI.h>
#include <Adafruit_PN532.h>

//Define the I2C pins
#define PN532_IRQ   (2)
#define PN532_RESET (3)   // Not connected by default on the NFC Shield
//Connect the shield I2C connection:
Adafruit_PN532 nfc(PN532_IRQ, PN532_RESET);

void setup(void) {
  Serial.begin(115200);
  Serial.println("Starting the NFC shield");
  nfc.begin();
  //Check the firmware version of the shield
  uint32_t versiondata = nfc.getFirmwareVersion();
  if (! versiondata) {
    Serial.print("Cant find the shield, check you connection");
    while (1); // wait
  }

  // Got ok data, print it out!
  Serial.println("Connected");
  //Print out the firmware version
  Serial.print("Firmware ver. ");
  Serial.print((versiondata>>16) & 0xFF, DEC);
  Serial.print('.'); Serial.println((versiondata>>8) & 0xFF, DEC);

  // Set the number of retry attempts to read from a card
  nfc.setPassiveActivationRetries(0xFF);

  // configure the board to read RFID tags
  nfc.SAMConfig();

  Serial.println("Waiting for a tag");
}
```

```
void loop(void) {
  //Decalre a variable to store state
  boolean success;
  //Buffer to store the ID of the card
  byte uid[] = { 0, 0, 0, 0, 0, 0, 0 };
  //Keychain that stores the card information
    byte keyID[] = {39, 246, 64, 175};
  int counter=0;
  //Variable to store the length of the ID
  byte uidLength;

  // Waits for and tag and reads the lenght of the ID and the ID
  success = nfc.readPassiveTargetID(PN532_MIFARE_ISO14443A,
  &uid[0], &uidLength);

  if (success) {

    for (uint8_t i=0; i < uidLength; i++)
    {
      if(uid[i]==keyID[i]){
        counter++;
      }
    }
    Serial.println();
    //If 4 out of 4 i right we unlock
    if(counter==4){
      Serial.println("unlocked");
      //if not the card is wrong
    }else{
        Serial.println("wrong card try again");
    }
    // Wait 1 second before continuing
    delay(1000);
    }
}
```

In the preceding code, we are basically doing the same reading operation as before, but instead of printing the ID number, we check it with the key code stored in the sketch. If the numbers match the one we stored, we will get a message in the serial monitor that says, unlocked.

Now when we have the basic functionality of the NFC reader covered in code, let's take a look at how to operate a servomotor, which we will be using to twist the lock on the physical door.

Connecting the motor

In order to actually open a door, we will need some help from the inside. **Servomotors** are the most common motors used in robotics and they are very easy to operate. There are two kinds of servomotors, there are the standard ones that only rotate 180 degrees and there are the continuous rotation servos that can rotate as much as you want in both directions. The trade-off with continuous rotation servos is that they are less precise. For this project, we will be using a standard servo since we only need to turn 45 degrees and the standard servo is usually more precise than the continuous rotation servo. Again, this depends on the kind of door you have. There is no such thing as a universal door, and you might need to modify the design of this project a bit depending on the door you have. In the next section of this chapter, we will take a closer look at the doors I used for this project, but first let's just take a quick look at how to operate the motor.

As servomotors are so common and popular to use among the Arduino community, the Arduino IDE has built-in functions for operating them. First, we need to connect the motor; they always come with three wires. In some cases, these wires are red, black and white or red, black and yellow in color. The red wires connect to **5+** and the black ones connect to **GND**. The yellow or white is the signal cable, which connects to a digital pin on the Arduino board, which in our case will be pin **9**. *Figure 6.4* shows how to hook the motor up to the Arduino:

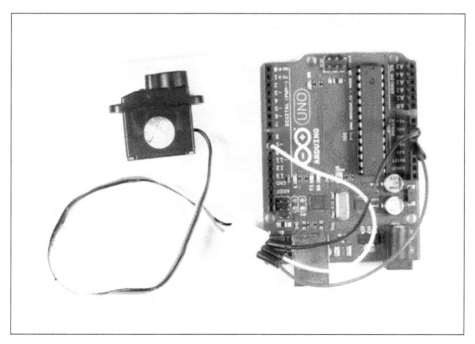

Figure 6.4: The servomotor connected to the Arduino Uno

Once you are hooked up and ready to go, upload the following code. If everything is connected right, the motor should start moving from its original start position to 180 degrees and back again. There are different sizes of servomotor available and the one used in this chapter is on the smaller side of the range. But even if it's small, it packs some serious torque, which was enough the turn the lock on my door. If it's too weak for your door, you could try a bigger one. Keep in mind that some of the bigger sized motors can't be powered straight from the Arduino board, since the current in the 5 V pin is not enough for some motors.

```
#include <Servo.h>
//Create servo object
Servo myservo;

void setup()
{
   //Name the signal pin
   myservo.attach(9);
}


void loop()
{
    //Rotate the motor from position 0 to 180 degrees
    myservo.write(0);
    delay(1000);
    myservo.write(180);
    delay(1000);
}
```

Putting the pieces together

Now, it's time for some serious construction in order to actually fit our project to the door. Mind you, this project was based on a standard lock used in Sweden; if you have a different door lock, this design might need some heavy modification. The goal was to keep the design of the construction as simple and transparent for this reason. I would suggest that you start by taking a look at *Figure 6.8* to get a general idea of the finished result before you get started.

In *Figure 6.5*, you will find the design measurement for the construction that holds the motor, shield, and Arduino board in place:

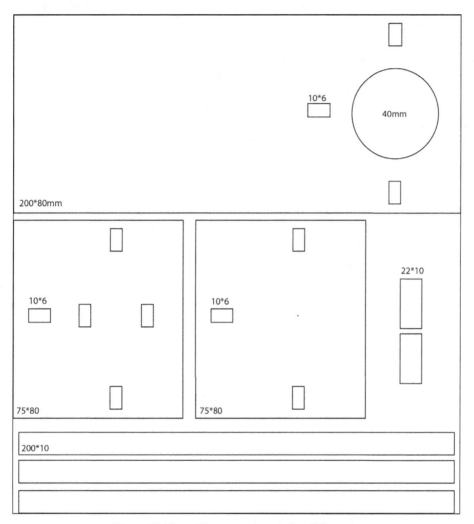

Figure 6.5: The cutting measurements for all the parts

The largest piece, which measures 200 x 80 mm, will act as our baseplate. The idea is that the front of the lock will be removed and then the baseplate will be inserted on the door. The front of the lock will then be put back into place, which will hold the baseplate in place. In this way, you don't have to modify the door in anyway or drill screws into the door to keep it in place. Drilling screws into the door is an easier option but most door locks can easily be unscrewed from the inside, so I do not recommend that you parentally fix anything to the door since these might be expensive to replace. The 75 x 80 mm piece to the left in *Figure 6.5* is where the motor will be place. The 22 x 10 mm pieces will be inserted into the middle holes of this piece in order to hold the motor in place. *Figure 6.6* should give you an idea of what this will look like. The second 75 x 80 mm piece to the right of the first one will only be used for steadiness. Both the 75 x 80 mm pieces and the 200 x 80 mm have three identical holes, which are placed in the same location. The idea is to run the three 200 x 10 mm pieces in order to make "floors" in the construction and hold everything together, which is also shown in *Figure 6.5*. The 200 x 10 mm pieces are longer than needed, but this will give you room to move things around if needed to get the right distance between the lock and motor. Once you are done, you can cut them into the right size.

Note that this design is based on a material that has a thickness of 6 mm. If you use any other material thickness, the pieces will not fit the holes. For this project, I used MDF since it's affordable and easy to use with a laser cutter. Even if it is possible to cut all the pieces by hand, I would recommend using a laser cutter for this project. If you don't have access to one, I would recommend that you use the design as a general guideline for how it can be done, but try to rethink the design so it fits with the tools and material available to you.

Figure 6.6: The motor plate in place together with the base plate

The measurement for the "claw" in *Figure 6.7* is 40 x 40 mm on all three pieces. However, I recommend that you make a few of them to try out with your own lock. The general guideline is that you don't want a snug fit between the lock and the "claw," since then, the axel of the motor needs to be perfectly aligned with the locks rotational axel. If you give it some room, the alignment does not need to be perfect. This makes the rest of the construction easier. *Figure 6.7* shows the "claw" that will be attached to the servo motor and placed over the actual lock:

Figure 6.7: The "claw" attached to the servomotor

The motor can be attached using either the screw mount on the actual motor or using plastic straps. I choose to use plastic straps, since again this gives the motor some room to move a little bit in case the axels don't align perfectly. But make sure not to strap it in too loose so that the motor falls out. Servomotors usually come with some different attachments for connecting them to other things and in *Figure 6.7* you can see the type of leaver I used for this project. This was simply screwed into the bottom of the "claw."

The process for putting everything together, I recommend, starts with attaching the baseplate to the door. Place the motor plate in the right position so it covers the lock, and make sure it's able to twist the lock. Then attach the last plate if some extra sturdiness is needed, and hopefully you'll end up with something that looks like *Figure 6.8*. The Arduino and shield is held into place using straps, since this is safer than attaching it with screws. You can use the screw holes on the NFC shield, but watch out so you don't scratch the PCB since this might cut one of the PCB connections. For this project, I cut the JST connector off the motor cable and soldered them to the NFC shield, which has an extension pin for all the pins on the Arduino board. The connections are the same as used before where the signal pin is connected to digital pin number **9** and the black cable to **GND** and the red one to 5V:

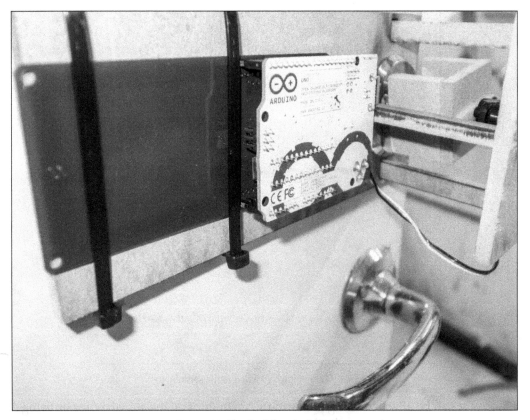

Figure 6.8: The final door lock in place

The final code

The following sketch is basically a combination of the three previous sketch examples in this chapter. First, it will activate the NFC shield and look for a tag or card. If the right card or tag is registered, this will activate the motor, which will turn the look and then the door will be open. The door will be open for 20 seconds and then the motor will turn back again locking the door. The code should work as supposed to, but the construction might need some tinkering before everything lines up as it should. Once everything is in place, it's time to upload the final code, which you will find here:

```
#include <Wire.h>
#include <SPI.h>
#include <Adafruit_PN532.h>
#include <Servo.h>

// Define the I2C pins
#define PN532_IRQ    (2)
#define PN532_RESET (3)   // Not connected by default on the NFC Shield
//Connect the shield I2C connection:
Adafruit_PN532 nfc(PN532_IRQ, PN532_RESET);
//Create servo object
Servo myservo;


void setup(void) {
  Serial.begin(115200);
  Serial.println("Starting the NFC shield");
  nfc.begin();
  //Check the firmware version of the shield
  uint32_t versiondata = nfc.getFirmwareVersion();
  if (! versiondata) {
    Serial.print("Cant find the shield, check you connection");
    while (1); // wait
   //Name the signal pin
 myservo.attach(9);
    myservo.write(0);
    delay(1000);


  }

  // Got ok data, print it out!
  Serial.println("Connected");
  //Print out the firmware version
```

```
    Serial.print("Firmware ver. ");
    Serial.print((versiondata>>16) & 0xFF, DEC);
    Serial.print('.'); Serial.println((versiondata>>8) & 0xFF, DEC);

    // Set the number of retry attempts to read from a card
    nfc.setPassiveActivationRetries(0xFF);

    // configure the board to read RFID tags
    nfc.SAMConfig();

    Serial.println("Waiting for a tag");
}

void loop(void) {
  //Decalre a variable to store state
  boolean success;
  //Buffer to store the ID of the card
  byte uid[] = { 0, 0, 0, 0, 0, 0, 0 };
  //Keychain that stores the card information
  byte keyID[] = {39, 246, 64, 175};
  int counter=0;
  //Variable to store the length of the ID
  byte uidLength;

// Waits for and tag and reads the length of the package and the //ID
  success = nfc.readPassiveTargetID(PN532_MIFARE_ISO14443A, &uid[0],
&uidLength);

  if (success) {

    for (uint8_t i=0; i < uidLength; i++)
    {
      if(uid[i]==keyID[i]){
       counter++;
      }

    }
    Serial.println();
    //If 4 out of 4 i right we unlock
    if(counter==4){
      Serial.println("Open door");
      myservo.write(180);
      //Keep the door open for 20sec
      delay(2000);
```

```
        Serial.println("Close door");
        myservo.write(0);
      //if not the card is wrong
    }else{
        Serial.println("wrong card try again");
    }
  // Wait 1 second before continuing
  delay(1000);
  }


}
```

Have a look at my NFC ring:

Figure 6.9: The wooden NFC ring

Wrapping things up

The last thing you have to do for your project is to choose a power source. Powering your board with a transformer is the most secure approach, since if the lock runs out of power it will not work. With most servos you could still turn the lock with a regular key, so don't worry. But with a transformer, you will need a cable from the lock to a wall socket, which might not be aesthetically pleasing for some. In that case, you can power the lock with a battery but keep in mind that the battery will eventually drain. Any transformer that outputs between 7–12V DC will probably do the trick and the same goes for batteries. The power source you choose needs to have a DC-Jack male connector to fit the Arduino board and do not forget to check the polarity of the DC-Jack. For most batteries, there are battery connectors you can buy; if you have the parts you can solder your own. You can't, however, power the lock via USB since there is not enough power for both the servo and NFC shield at the same time.

As I said before, NFC cards and tags are available in all different shapes and sizes. There are even NFC tags made to be inserted into the body. I actually have two friends that have NFC implants, which technically makes them cyborgs. But if you're not on the extreme edge of wearables, I would hold off on implants and stick with tags for now. There is still tons of fun to be had making your own gear and hiding NFC tags in them. For inspiration, you will find my NFC ring in *Figure 6.9*. I simply carved it out of wood and fitted an NFC tag in the front. I wanted to go for a low/high tech look with the antenna and chip visible in the front.

Summary

In this chapter, you learned a bit about NFC technology and how it can be used as a means to give objects a unique identifier that can be digitally registered. You also learned a bit about servomotors and how they can be used to move things around. The construction part of this project might seem tricky to some and it is but with some patience, I am sure you will get there. The good part is that once you're done with the door, you can have a lot of fun creating new objects to open the door with minimum effort. Instead of hiding a spare key under the doormat or a garden stone, why not make the actual doormat or stone into the key. I bet no one would ever think of trying to open a door with these objects. If you are worried about losing your tags, there are re-programmable tags available, which means you can clone your tags.

In the next chapter, we will stick with wireless communication but we will be switching over to Bluetooth. This gives us a larger range of functions to play around with.

7
Hands-on BLE

In this chapter, we will take a look at Bluetooth technology and how to make your own activity and gesture-monitoring device. Bluetooth is a standard for wireless communication between devices that was developed by the Swedish telecom company Ericsson in the 90s. The Nordic Viking king, Harald Blåtand, who united the Norwegians and the Danish, was known for his communication skills and as a great speaker inspired the name Bluetooth. The word Bluetooth is a direct translation of his last name, which seemed fitting.

The actual technology was off to a slow start and struggled for its existence for the first year, but now it's safe to say that it has become a success and the market has exploded with Bluetooth connected devices. Today, it seems that everything is connected with Bluetooth. Cars, controllers, headphones, keyboards, and even lamps connect to other devices over Bluetooth. In 2010, Sony started the development of a Bluetooth version called **smart Bluetooth**, a smaller low-powered version of Bluetooth, which targeted the market of fitness and healthcare. As in the case of all wearables, size and power composition are always a concern, so smart Bluetooth needed to be small and power efficient. However, small and power efficiency is loved by all, so a part of the Sony development became a part of the Bluetooth version 4.0 standard. This standard is also known as **BLE**, which stands for **Bluetooth low energy**, and as I say, "With low energy comes great possibilities".

BLE will act as the base for this chapter where we will be using the technology to create our own gesture-tracking device. As always, the projects in this book aim to get you started, then get you to take over and make the project your own. Maybe you want to make a keyboardless keyboard, a gesture controlled lamp, or add some cool sound effect to your karate training. The possibilities are endless.

In this chapter, we will also introduce a new prototype board called the **Blend Micro**, which is a very special board, prefect for creating wearables.

Materials needed:

- Blend Micro board
- ADXL355 accelerometer
- Mini or regular breadboard
- Mobile phone supporting Bluetooth Version 4.0
- Cables
- 3.7V 150–500mAh lithium Battery
- 220Ω resistor
- Micro USB cable

Hello Blend Micro

The Blend Micro board produced by RedBearLabs combines the ATmega32U4 chip used in many Arduino boards with the Nordic nRF8001 Bluetooth chips. The boards are very small, but still feature the same functions as a normal Arduino board with the added feature of BLE. This board can connect to anything that supports Bluetooth version 4.0 and above.

In order to be able to program the board from the Arduino IDE, we need to add support for this type of board. In the Arduino IDE 1.6.4 and above, adding support has been made very easy. If you take a look in the menu by navigating to **Tools | Boards**, at the top of the list you will find the **Boards Manager...**. If you open it up, it should look something like *Figure 7.1*:

Figure 7.1: The Boards Manager

Just search for Blend Micro and press the **Install** button and all the necessary files to support the Blend Micro board will be installed. If you are using an older version of the IDE, you will need to manually modify the IDE according to the following steps:

1. First, head over to the Red Bear GitHub on the following link and download the `.zip` file from `https://github.com/RedBearLab/Blend/RL`.

2. Unzip the file once downloaded and locate the hardware folder inside `Blend-****` | `Arduino`.

3. Copy this folder and move to the Arduino folder created inside your documents folder where the IDE is installed. Place the copied hardware folder into the `Documents` | `Arduino` folder. It should end up in the same level as your libraries folder.

In the next step, we need to modify the `main.cpp` file located in the Arduino IDE application folder in order for the Blend Micro to be listed in the IDE. Note that this is not the same folder as the `Documents` | `Arduino` folder. The folder we are looking for is inside the folder where you find your Arduino application.

On a windows computer, you will find the `main.cpp` folder:

1. Find the `Arduino IDE` folder.

2. Navigate to `hardware` | `arduino` | `cores` | `arduino` | `main.cpp`.

On an OS X computer, you need to:

1. Find your Arduino application and then right-click on the application.

2. Choose **Show package content**, which will open up a folder in finder.

3. Navigate through the `Contents` | `Resources` | `Java` | `hardware` | `arduino` | `cores` | `arduino` | `main.cpp` folder.

Once you have located the `main.cpp` file, open it with any text editor and edit the code with the following code and then save it:

```
#if defined(BLEND_MICRO_8MHZ)
PPLCRS |= 0x  10;
while (!(PPLCRS & (1<<PLOCK)));
#elif defined(BLEND_MICRO_16MHZ)
run (i.e. overclock running at 3.3v)
CLKPR = 0x80;
CLKPR = 0;
#endif
```

If you have your IDE open, you will need to restart the IDE in order for the changes to take effect. Once reloaded, check your board menu by navigating to **Tools | Board**, and you should be able to find the Blend Micro in the list. As you may have noticed, there are two different Blend Micro boards, one 8MHz and one 16MHz. The Blend Micro runs at 8MHz by default, but you can run it at a higher processing speed of 16MHz if you want. The risk is that the board might be a bit less stable at the higher speed even if this risk is low.

Once we have added the support for the board, we need to install the libraries for the BLE chip. In Arduino 1.6.4 and above, libraries are installed through library manager. If you are working with an older version of the IDE or if the library is not available through the manager, you can still install it manually. Download the following library `.zip` files from the following links:

`https://github.com/NordicSemiconductor/ble-sdk-arduino/`

`https://github.com/RedBearLab/nRF8001.`

Once you've downloaded, perform the following:

1. Unzip both `.zip` files and open up you IDE.
2. Navigate to the **Sketch | Add library** menu.
3. Select the libraries folder from the downloaded `nRF8001` folder `nRF8001-****` | `Arduino` | `libraries` and import it.
4. Do the same for the `ble_sdk_arduino` | `libraries` folder.

If you are running a Windows computer, you will need to install an additional driver for the board. In the following link on the Red Bear website, you will find a short tutorial on how to do this and where to download the drivers from:

`http://redbearlab.com/windows-driver/`

In order to test that everything works, upload the standard blink example found under **File | Example | Basic | Blink** or write the following code in the IDE and upload it to your board:

```
void setup() {

  pinMode(13, OUTPUT);
}
```

```
void loop() {
  digitalWrite(13, HIGH);
  delay(1000);
  digitalWrite(13, LOW);
  delay(1000);
}
```

If everything was installed correctly, the on-board LED on the Blend Micro should start to blink with a 1 second delay. If it does not, first make sure that:

- The `main.cpp` was changed properly and saved
- The board is connected
- The right board and com port is selected in the tools menu
- The IDE was restarted after changing the `main.cpp` file

When you're done with setting up the IDE, you can play around a bit by controlling the Blend Micro from a mobile phone.

The Blend Micro app

Red Bear has developed an application for both Android and IOS, which are great companions for developing and debugging the Blend boards. Start by downloading the application to you mobile phone. If you are running Android, you will find the application at `https://play.google.com/store/apps/details?id=com.redbear.redbearbleclient`.

And for IOS:

`https://itunes.apple.com/app/ble-controller/id855062200`.

You can also search for BLE Controller in the Android Play store or in iTunes.

In order to give some feedback on the Blend Micro let's set it up on a breadboard with an external LED, which we will control from the mobile application. When you buy a Blend Micro board, it comes with the male pin headers unsoldered. I suggest you keep them unsoldered since we want to keep the circuit for the project as flat as possible later on. For this next part, you can just place the male pin headers on a breadboard so that they line up with the pins of the Blend Micro, and then place the Blend Micro on top as shown in *Figure 7.2*. In this way, we can remove it later on. This might create a bit of a glitch sometimes, so make sure the pins are pressed snug against the solder pads:

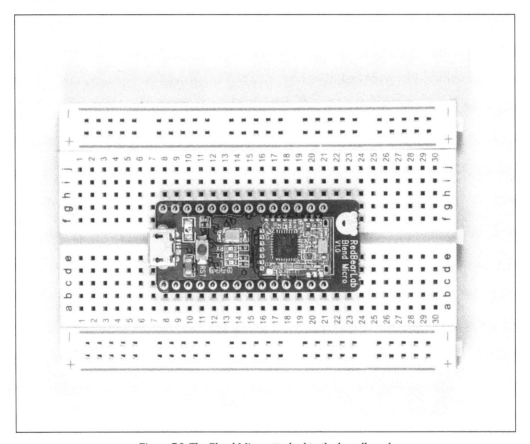

Figure 7.2: The Blend Micro attached to the breadboard

Once the Blend Micro is placed on the breadboard, we can add the external LED and we will need a 220Ω resistor as well. *Figure 7.3* shows everything connected;

1. Connect the resistor between **GND** and the ground line.

2. Connect the positive leg of the LED to pin **3** and the negative to the ground line.

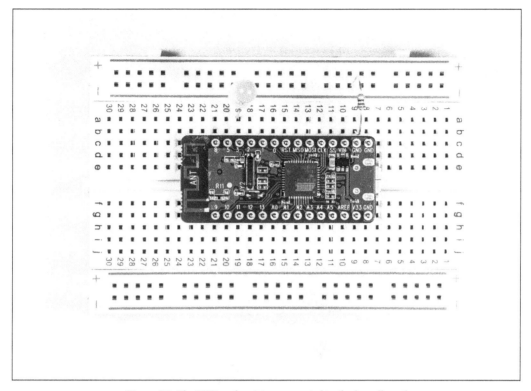

Figure 7.3: The LED and resistor connected to the breadboard

Note that I flipped the Blend Micro over in *Figure 7.3*. The only reason for this is to expose the numbering of pins on the board.

Once your breadboard and Blend Micro is set up, we need to upload some code. At this point, we just want to make sure that things are working, so to make things easy, we will be using some of the examples from the BLE library:

1. Navigate to **File | Examles | BLE | BLEcontrollerSketch**.
2. Select the **Blend Micro** in the board menu.
3. Check that the right USB port is selected under **Tools**.
4. Upload the code.

This sketch acts as a firmware for the Blend Micro, so we can access the basic functions from the mobile application. Now open up the mobile application and press on the **Menu** button to open the menu as shown in *Figure 7.4*:

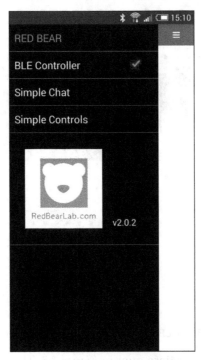

Figure 7.4: The BLE Controller app menu

Choose the BLE Controller and you will be prompted with a screen and **Scan** button. Press the **Scan** button and the application will start searching for BLE devices in your vicinity. If the Blend Micro is powered, it should find it within a few seconds as shown in *Figure 7.5*:

Figure 7.5: Scanning for BLE devices

Figure 7.5 shows the expected result of the scan. It shows the local name (which is Blend Micro) and the actual MAC address of the BLE chips, which is a unique address for your particular chip and board. Press the arrow button and wait a few seconds for the application to pair with you board. Once paired, the interface as show in *Figure 7.6* should appear:

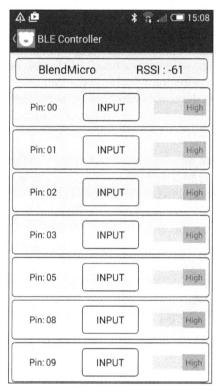

Figure 7.6: The options for direct pin control

From the application, you can set the basic pin modes and basic state options. In order to try your breadboard setup, switch the pin mode state on pin **3** from **Input to Output** and then toggle the **Low** to **High** switch. If everything is connected, the LED should turn on and off. The **PWM** and **Servo** options are also included, so you can also try to switch the pin mode to **PWM** and the **Low/High** toggle will switch to a slider. Sliding the slider around should change the intensity of the LED light. Note that there might be a slight delay because the application has to send the information to the Blend Micro board.

Now that we know that the Blend Micro board is up and running, we can set up a connection to another device. Let's take a look at how to connect an accelerometer to the mix and get started with coding some algorithms for tracking gestures.

Gesture tracking

For this chapter, I will be using an accelerometer breakout board based on the **ADXL335** chip. You can basically use any accelerometer you find for this project. If you already have gone through this chapter, you might have the combined accelerometer/compass/gyro from earlier, which will also work. The reason for using the ADXL335 in this chapter is that it has a different form factor, which fits this project better. This accelerometer also makes readout on analog pins, so there is no need for I2C as presented in *Chapter 3*, *Bike Gloves*. Accelerometers are suited for measuring acceleration of gravity or as a result of motion of chocks, however, they are known to be very noisy sensors, which means that their readouts are not always exact. The following sketch demonstrates how to "smoothen" the readouts and how to display the information in degrees, as well as indicate which direction the accelerometer is moving.

First, we need to connect our accelerometer to the Blend Micro board. I recommend connecting it to the small breadboard we used before while connecting the LED to the Blend Micro. Some accelerometers breakout boards come with the pins unsoldered, so before you connect it to the breadboard, you might need to solder them in place or just place it on the pin headers on the breadboard as done with the Blend Micro:

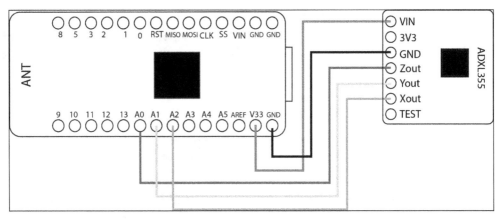

Figure 7.7: The ADXL355 connected to the Blend Micro

You can connect it as per the following sequence as shown in *Figure 7.7*:

- **VIN** connects to **V33**
- **Xout** to analogpin **A2**
- **Yout** to analogpin **A1**
- **Zout** to analogpin A0
- **GND** to **GND**

Figure 7.7 shows all the connections that need to be made. In the case of the ADXL335, you can connect it to either 3.3V or 5V if you ever use it with a standard Arduino.

Getting good values from an accelerometer is a science on its own, and unfortunately, this chapter is not long enough to really get into depth on accelerometers. To get you started, we will take a look at how to detect direction and degrees of movement.

Let's try out the accelerometer with the Blend Micro and see if we can get some values that are good for most projects. Even if accelerometers are noisy sensors, they still work for most applications based on human interaction. Even if the accelerometer generates a faulty value readout in-between moving it 0.01 mm and 0.02 mm, it's very hard to move the sensor as little as this. In most cases, human interactions have a greater span and a super exact sensor makes little difference since humans have a hard time being perfectly still.

As you will see in the following sketch, we can still get pretty decent measurements in degrees with some simple code. Upload the following sketch to your board and open up your serial monitor.

```
int samples=50;
int acc=20;
float sampleX;
float sampleY;
float sampleZ;
float lastX=0;
float lastY=0;
float lastZ=0;
float degreeX=0;
float degreeY=0;
float degreeZ=0;
float zeroX;
float zeroY;
float zeroZ;
void setup(){
  //Initiate the serial communication
  Serial.begin(57600);
  //Calculate the zero position
  zeroX=calibrateX();
  zeroY=calibrateY();
  zeroZ=calibrateZ();
}

void loop(){
  //Get the sensor data for x, y and z axis and store it
```

```
  float x=calibrateX();
  float y=calibrateY();
  float z=calibrateZ();
  /*Calculate the degrees by subtracting the zero position for the
recent readout*/
  degreeX=x-zeroX;
  degreeY=y-zeroY;
  degreeZ=z-zeroZ;
/*Delay only needed when printing information to the serial monitor*/
  delay(200);
  Serial.println(degreeX);
  /*If the axis is smaller or bigger then the value we use for
accuracy we print out the data and degrees*/
  if(x<lastX-acc){
    Serial.print("X Degrees: ");
    Serial.print(degreeX);
    Serial.println(" Moving right");
  }
  if(x>lastX+acc){
    Serial.print("X Degrees: ");
    Serial.print(degreeX);
    Serial.println("Moving left");
  }
  if(y<lastY-acc){
    Serial.print("Y Degrees: ");
    Serial.print(degreeY);
    Serial.println("Moving backwards");
  }
  if(y>lastY+acc){
    Serial.print("Y Degrees: ");
    Serial.print(degreeY);
    Serial.println("Moving forwards");
  }
  if(z<lastZ-acc){
    Serial.print("Z Degrees: ");
    Serial.print(degreeZ);
    Serial.println("Moving up");
  }
  if(z>lastZ+acc){
    Serial.print("Z Degrees: ");
    Serial.print(degreeZ);
    Serial.println("Moving down");
  }
  lastX=x;
```

```
    lastY=y;
    lastZ=z;
}
/*Functions for better readouts which takes 50 samples and calculate
the median*/
void calibrateX(){

  for(int i=0;i<samples;i++){
     sampleX += analogRead(2);
  }
  sampleX = sampleX/samples;
}
void calibrateY(){

  for(int i=0;i<samples;i++){
     sampleY += analogRead(1);
  }
  sampleY = sampleY/samples;
}
void calibrateZ(){

  for(int i=0;i<samples;i++){
     sampleZ += analogRead(1);
  }
  sampleZ = sampleZ/samples;
}
```

In the preceding sketch, we are performing two operations, calculating the degrees of movement and detecting if the movement is above or below the desired range. As I said before, accelerometers are noisy, which means that the values jump around a bit even if you don't move the sensor. This is why we implemented a threshold and if you want a higher or lower accuracy, you can change the acc variable. Again, since the sensor is a bit noisy, we implemented a function for readouts on the axes. The function takes 50 sample readings and calculates the average from these readings in order to get smoother readouts. The median is returned every time we call each of the functions.

Next, we will take a look at how we can send information over the Bluetooth instead. Since some of you might have a computer which doesn't support Bluetooth Version 4.0, we will use a mobile phone to act as our serial port. With the Blend Micro, we can only send one byte at the time, so this sketch shows how messages can be formatted so we can send full messages.

Start by uploading the following code to your board:

```
#include <SPI.h>
#include <EEPROM.h>
#include <boards.h>
#include <RBL_nRF8001.h>
unsigned char buf[16] = {
  0};
unsigned char len = 0;

int samples=50;
float sampleX;
float sampleY;
float sampleZ;
int acc=20;
float lastX=0;
float lastY=0;
float lastZ=0;
void setup(){
  // Init. and start BLE library.
  ble_begin();
  Serial.begin(57600);
}

void loop(){

  float x=calibrateX();
  float y=calibrateY();
  float z=calibrateZ();

  if ( ble_connected() ){
    if(x<lastX-acc){
      ble_write('R');
      ble_write('i');
      ble_write('g');
      ble_write('h');
      ble_write('t');
      ble_write(' ');
    }
    if(x>lastX+acc){
      ble_write('L');
      ble_write('e');
      ble_write('f');
```

```
        ble_write('t');
        ble_write(' ');
      }
      if(y<lastY-acc){
        ble_write('B');
        ble_write('a');
        ble_write('c');
        ble_write('k');
        ble_write(' ');
      }
      if(y>lastY+acc){
        ble_write('F');
        ble_write('o');
        ble_write('r');
        ble_write('w');
        ble_write('a');
        ble_write('r');
        ble_write('d');
        ble_write(' ');
      }
      if(z<lastZ-acc){
        ble_write('U');
        ble_write('p');
        ble_write(' ');
      }
      if(z>lastZ+acc){
        ble_write('D');
        ble_write('o');
        ble_write('w');
        ble_write('n');
        ble_write(' ');
      }
    }
    ble_do_events();
    lastX=x;
    lastY=y;
    lastZ=z;
}

float calibrateX(){

  for(int i=0;i<samples;i++){
    sampleX += analogRead(2);
```

```
    }
    sampleX = sampleX/samples;

    return sampleX;
  }
  float calibrateY(){

    for(int i=0;i<samples;i++){
      sampleY += analogRead(1);
    }
    sampleY = sampleY/samples;

    return sampleY;
  }
  float calibrateZ(){

    for(int i=0;i<samples;i++){
      sampleZ += analogRead(0);
    }
    sampleZ = sampleZ/samples;

    return sampleZ;
  }
```

Once the code is uploaded to the board, it is time to switch over to your mobile phone and open the BLE Controller app you previously used in this chapter. Open up the app and select **Simple chat** in the menu as shown in *Figure 7.4*. Then scan for devices and select your **BlendMicro**. This will open up a new window that looks like *Figure 7.8*, and as soon as you start moving your accelerometer around, the different directions should appear on your mobile screen:

Figure7.8: The app printout from the accelerometer

Wrapping things up

Once you have gone through the setup process in this chapter and tested a working connection between your Blend Micro and another device, you can move on from the breadboard and solder the connections shown in *Figure 7.6* with wires. You also need to add a connector to **VIN** and **GND** for a battery. This depends on the type of battery you want to connect, but if it's a small lithium one, there is a chance that you will need a two-pin female JST connector.

The gesture-tracking device created in this chapter can have a range of functions, depending on where you put it. In some of these cases, weatherproofing might be of importance but making a waterproof casing from scratch is trickier than might be expected. In my case, I found that an old film bottle works perfectly as a container for the Blend Micro and accelerometer used in this project. It protects the electronics from any liquids and is still small enough to place into other objects. Since not everyone is into film photography these days, any plastic container with a good seal will do. Once inside a container, you can start experimenting with the circuit placing it on different parts of your body, in order to figure out what type of data you might want to collect. Once you know, you can start experimenting with different garments to place them in. The best way is to find a material that is water resistant and place your electronics in that. Then place them inside a garment either by sewing it into place or using Velcro so that things can easily be removed. If you are skilled in sewing, an option is to look at water resistant fabrics that can be used to sew a pouch into a garment to hold the electronics in place.

As I mentioned in the beginning of this book, the possibilities are endless with Bluetooth and I can probably write an entire book about Arduino and Bluetooth. In order to really dig into Bluetooth possibilities, eventually we will need to turn to some mobile application development, but unfortunately there is no room for that in this book. There are also a few things on other wearable technology I would like to cover before the book ends.

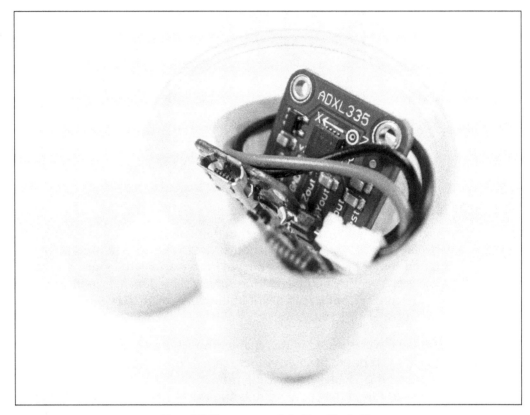

Figure 7.9: The components inside a film bottle

If you want to learn more about the possibilities of Blend Micro, both the nRF8001 and Arduino SDK libraries used in this chapter have a few examples included that are worth mentioning as inspiration for further development. If you want to learn more about accelerometer, I recommend you head over to Jeff Rowbergs website about "**Keyglove**". Keyglove is an open source project about a connected glove, which can be used for a range of applications and the website has a good segment about accelerometers.

```
http://www.keyglove.net
```

If you want to learn more about developing mobile applications that can be connected to the Blend Micro the BLE Controller app used in this chapter is available as open source together with the SDK for Android and IOS.

```
https://github.com/RedBearLab/Android
```

```
https://github.com/RedBearLab/IOS
```

If you are completely new to mobile application development, but want to learn more about it in a wearable context, I recommend the book *Professional Android Wearables, Wrox Press* by *David Cuartielles Ruiz* and *Andreas Goransson*.

Summary

This chapter has presented the basics of getting you up and running with Bluetooth. In this chapter, we introduced BLE with the Blend Micro and had a closer look at how to get stared with gesture recognition using accelerometers. We also learned two ways of adding support for third party boards to the Arduino IDE.

In the next chapter, we will introduce yet another Arduino compatible board using wireless communication called the **Particle core**, but this time, the focus will be on Wi-Fi. The next chapter is an "in-between" chapter where we will introduce the Spark Core. This will act as the foundation for the last chapter and project where we will combine most of the technologies and knowledge from the previous chapters.

8
On the Wi-fly

In this chapter, we will have a look at yet another new Arduino compatible board called the **Particle Core**. The big feature with this board is that it includes Wi-Fi connectivity in a very small form factor. I guess most of you are familiar with Wi-Fi technology these days as Wi-Fi has become a standard for wireless communication in almost all personal computers and smart phones, and in recent years we can even find Wi-Fi technology in many consumer electronics. One of the first consumer products including Wi-Fi technology which was not a personal computer was weirdly a bathroom scale. Today, we find stereos, robot vacuum cleaners, light bulbs, and even refrigerators that include Wi-Fi technology. The reason for this is of course to connect these products to the Internet.

Even though I am not that old, I still remember a time when anything connected to the Internet needed a cable, but fortunately those days are long gone. Wi-Fi technology is still older than one might think. The development started in the early 70s, but it was not until the late 90s that Wi-Fi started to become commercially available and used. Even in the beginning of the age of Wi-Fi for personal computers, Internet connections over cable were preferred since the speed over Wi-Fi nowhere near matched the speed over cable connections. Wi-Fi has still not caught up with cable connections, but it is still fast enough for our everyday needs.

Wi-Fi is a local area wireless computer networking technology, which can be used for peer-to-peer connections, or connecting to the Internet through a router. The actual name of the technology is **IEEE 802.11x** where x is a letter that indicates the version of the technology. The term Wi-Fi was not used until the late 90s when the name was rebranded for commercial purposes. The reason was simple: it is easier to say Wi-Fi than to refer to the technology as IEEE 802.11. In recent years, Wi-Fi has become faster, smaller, and cheaper, which has made it one of the backbone technologies of the "Internet of Things".

By now, you are familiar with wearables and some consider wearables as a sub-domain in the Internet of Things where ordinary objects are modified with computational abilities and connected to the Internet. Personally, I like to think that wearables will become the "glue" devices that connect all of these things to the Internet.

So, of course we will have a look at Wi-Fi technology in this book since I find it fitting to end the book with what will become the future (hopefully the future will prove me right). The remaining two chapters will focus on the Particle Core, where we will first introduce the board and how to wirelessly program it then, in the last chapter, use the Particle Core to make our own smart watch.

In this chapter, we will have a look at how to connect to the Core board and how to code for it. We will also have a look at how to send information to and from the Internet and how to control the board from a simple web page. We will also have a look at how to connect the Core to third-party services through the use of a mash-up service called **IFTTT**.

The following are the materials that you will need for this chapter:

- Particle Core
- Micro USB (included in the development kit)
- Mini breadboard (included in the development kit)
- LEDs
- A 220Ω resistor
- An Android or iPhone phone

The Particle Core

The Particle Core is an Arduino-compatible board but with some reservations. In fact, the Particle Core is based on ARM Cortex-M3 processors, which use a different architecture from the Atmega microprocessors used in most Arduino-compatible boards. This means that the Particle Core can't be programmed from the Arduino IDE at the moment, but don't worry as there are other options.

Still we can consider the Particle Core to be Arduino-familiar since we can program them using the same commands and structure as used to program all Arduino boards. Many of the libraries available for Arduino boards also work for the Particle Core and some are specifically developed for the Particle Core in order to utilize the Wi-Fi connectivity. *Figure 8.1* shows the development kit available from:

```
https://www.particle.io/.
```

This kit includes the Particle Core board, a mini breadboard, and a micro USB cable:

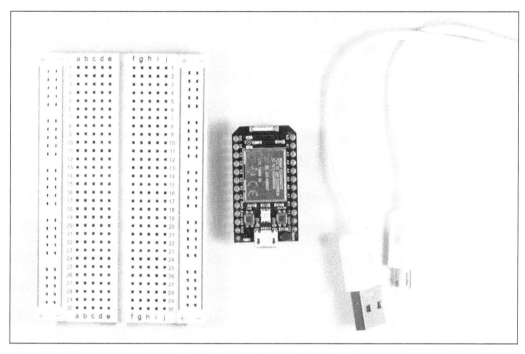

Figure 8.1: The Particle Core starter kit

In order to get started on programming the Particle Core, we need to connect it to your network. To do so, we will need some help from the Particle app in order to connect your Particle Core to your network. The app is available for both Android at `https://play.google.com/store/apps/details?id=io.particle.android.app` and for IOS at: `https://itunes.apple.com/us/app/particle-build-photon-electron/id991459054?ls=1&mt=8`.

To set up your Particle Core, perform the following steps:

1. Download and install the app on your smartphone.
2. Once installed, you need to create a new account or sign in with an existing one.
3. Power the Particle Core using the included Micro USB cable.
4. Connect the phone to the Wi-Fi network you want to use and then enter the password for the network in the app.

The app should automatically find your Particle Core board and connect it to the network. In order to debug the different modes of the Particle Core, the board includes an RGB LED on the top in order to indicate the different states:

- The blue light blinking means that the board is looking for Wi-Fi credentials
- The blue light full on means that the board is getting information from the app
- Green blinking means that the board is connecting to a Wi-Fi network
- Cyan blinking means that the board is connecting to the Particle cloud
- Magenta blinking means that the board is updating its firmware to the newest version
- Pulsating magenta means that the board is connected and ready to use

If the app should fail to connect and the LED just keeps blinking blue, try to connect one more time from the app. If the board gets stuck on green blinking, try holding down the small **Mode** button on top of the board for a few seconds until the light starts blinking blue again. The **Mode** button is located to the left of the RGB LED There is a button to the right of the LED also marked **RST**, which means RESET, and as suggested by its name, this button resets the program on the board.

Hopefully everything went smoothly and if so, the **Tinker** screen should appear in the app as shown in *Figure 8.2*. The Tinker screen enables basic read and write operations on the pins on the Particle Core board from the phone app:

Figure 8.2 The Tinker screen in the core app

In order to test that everything works:

- Try tapping on the **D7** pin in the application and set it to **digital write**
- Press on it again and set the mode to **High**. This should turn on a small blue LED at the top of the Particle Core next to the USB connector
- If the blue LED turns on, everything works and the Particle Core board is connected to your Wi-Fi network
- If not, run the connection process again and make sure you connect to the right network from the application

Once connected to the Particle Core, you also have the option to change the name of your board, which might be a good idea if you have multiple boards. If not, you can leave the name as is.

Programming for the Particle Core

As I mentioned before, we can't use the Arduino IDE to program the Particle Core board since the Arduino IDE has no support for this type of board at the time I am writing this book. But using the Arduino IDE to program the Particle Core is to miss of the point of using them a bit. Since these boards implement Wi-Fi technology, we can program them wirelessly, which might be a huge benefit in some cases and especially when it comes to wearable projects. Debugging a wearable project is a pain sometimes if you constantly have to be connected to a computer in order to make changes to the code. So wireless programming alone is a reason for considering the Particle Core for a wearable project. Then we also have a connection to the Internet, which makes for some really interesting projects.

In order to make these projects, we need to be able to program the Particle Core, and Particle, the company behind the Particle Core, has developed a web-based IDE that can be accessed at the following link: `https://build.particle.io/login`.

This link will direct you to a login screen where you log in with the account information used to create an account through the phone app. When you have logged in, it should look something like *Figure 8.3*:

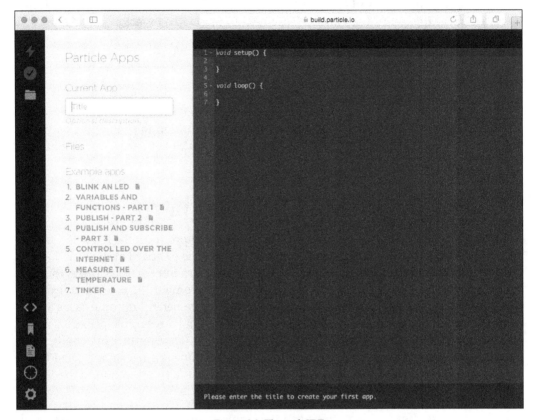

Figure 8.3: The web IDE

As you can tell from *Figure 8.3*, this IDE looks a bit different from the Arduino IDE but the basic functions remain the same. To the left you have your basic operations. First, at the top you will find a button marked with a small lightning, which is the flashing button that is equivalent to the upload button in the Arduino IDE. This button pushes the code from the web IDE to the Particle Core board. Since this is done over Wi-Fi, the flashing will not be as fast on a regular Arduino board. But usually, it does not take more than a few seconds. In order to test the flashing of code, add the following code to the web IDE in the code input area, which is the large dark grey space shown in *Figure 8.3*:

```
int led = D7;

void setup(){
```

```
pinMode(led,OUTPUT);
}

void loop(){
digitalWrite(led,HIGH);
delay(1000);
digitalWrite(led,LOW);
delay(1000);
}
```

As you might have guessed, the code will turn the blue LED next to the USB port on and off with a 1 second delay. Now try flashing the code to your Particle Core board. If you are on the same Wi-Fi, the IDE should automatically find your board and upload the code. The RGB LED should turn to cyan and start to blink when flashing the board. Once the flashing is done, the RGB LED should turn back to a pulsating magenta and the blue LED should start to blink.

Below the **Flash** button, you will find a **Verify** button and under this one, you will find a **Save** button. Both have the same function as in the Arduino IDE. A bit further down, you will find five more buttons. The first one hides and shows the code panel where you can find your saved sketches and recently used ones. Next you have the **Libraries** where you can search for all the available libraries. The **Docs** button will direct you to a tutorial page with more information about the Particle Core and available services. Below the **Docs** button, you will find the **Device** button, which shows/hides the available devices connected. The last button is the settings button, where you can change your password and find your access token, which might be needed for some services to connect to the Particle Core.

Figure 8.4: The IDE menu buttons

The Dashboard

So hopefully you have your Particle Core connected to your Wi-Fi network and have succeeded in flashing new code to the board over the air. Wireless programming has many benefits but there is more to the Particle Core, which makes it very interesting for wearable applications. What's the point of being connected to the Internet if we can't use it to send data? The people behind the Particle Core have already thought of this and have implemented some really nice features for data transfers and remote communication with the board.

First off, there is the Particle Dashboard, which you can find at the following link: `https://dashboard.particle.io`.

This will open up a website, which should look something like *Figure 8.5*. If you are logged in to the web IDE, the dashboard should automatically open. If not, you need to enter your account information, which is the same as for the app and web IDE:

Figure 8.5: The Particle Dashboard

This dashboard is able to visualize data sent from the Particle Core board. In software architecture, there is a messaging pattern often used called **publish-subscribe**, or **pub/sub** for short. The basic idea behind this form of messaging is that a sender of messages, called a **publisher**, can send messages without a specific receiver in mind. On the other side, you have subscribers who can collect data without the need to connect to the publisher. This works by naming the data and storing the data on a server where it will remain until changed or removed and the subscribers can collect the data from this server. The whole idea is similar to a radio where a publisher acts as a radio station and anyone with a radio receiver can tune in to a channel and listen to whatever is broadcasted on this channel.

In practice, this means that you can connect your Particle Core and broadcast whatever sensor data you want and anyone or any application can listen in to your channel and use this data for something else. In order to try this out for yourself, we first need to have a look at some code:

```
    void setup() {
  //No need to do anything
}

void loop() {
        //Publish the data
        Spark.publish("MyMessage","Hello",60,PUBLIC);
        //Wait for a bit
        delay(1000);
        //Publish the data
        Spark.publish("MyMessage","Goodby",60, PUBLIC);
        // Wait for a bit
        delay(1000);

}
```

In the code example just given, we are using the publishing command to send information to the dashboard. This first parameter of this function is the event name, or channel name if you like. If the event name does not exist, it will be created.

The next parameter is the actual data you want to send. In the case of the previous code, these are the messages `Hello` and `Goodbye`.

The third parameter is the TTL value, which is short for time-to-live, which limits the lifetime of data in a computer or network. TTL prevents data from circulating indefinitely. In the case of the Particle Core, the default TTL is 60 milliseconds and if you don't have a good reason to change it, I would just stick with this value.

The last parameter is used to indicate whether you want your data to be private or public, for example, if you want other people to be able to access your data or not. You do not need to add the last two parameters if you don't want to. If you don't, the default values will be used, which are TTL 60 and the data will be published as public.

So to sum up, what this sketch does is basically send a message every second, where it alternates between `Hello` and `Goodbye`. Flash the code from the web IDE and once done, open up your dashboard. It should start to look something like *Figure 8.6*:

Figure 8.6 Receiving data in the dashboard

As you can see, we have multiple messages displayed; every second, the data is updated since this is the delay we use in the code. We can also see when the data was published and from which device. In my case, I have changed the name of the Particle Core to `tonyC`. The interface also includes a visualization of a timeline when the data was received.

Now let's say we have the opposite scenario, where you want to read data from someone else's data stream. In order for the next code to work, you need an additional Particle Core set up with the first code example below and the second Particle Core with the second example code. I don't expect you to buy two Particle Cores, but you might have a friend that owns one who could help you to try out this next part. Or you could head over to the Particle community forum and ask someone to help you, or check whether someone has an open data stream you could try subscribing to: `https://community.particle.io/`.

Once you have found another Particle Core buddy, ask them to flash the following onto their board:

```
void setup() {
//Do nothing
}

void loop() {

Spark.publish("tonyC_data_stream","ON");
delay(2000);
Spark.publish("tonyC_data_stream","OFF");
delay(2000);

}
```

Make sure you change the event name to something that corresponds to your application. If everyone that reads this book uses the same event name, we might end up in trouble! So just make sure that the name is unique. The reason for this is that the name works as a filter. In the next code, we will search for this event name in the `subscribe` command. But this command is very smart so it uses the event name as a prefix filter. This means that if we were to add the event name `hello`, the command would subscribe to data streams that starts with `hello`, for example, streams named `helloworld`, `hello_all`, or `hello_to_me`. But if you make the event name unique enough, you will avoid unwanted data subscriptions.

Once you have one Particle Core set up with the previous code, flash the second one with the following code:

```
int led = D7;

void setup() {
pinMode(led,OUTPUT);
Particle.subscribe("tonyC_data_stream", doFunction);
}
```

```
void loop() {
//Do nothing

}

void doFunction(const char *event, const char *data)
{

    //Compares the incoming data to the string "ON"
  if (strcmp(data,"ON")==0) {
digitalWrite(led,LOW);
  }
    //Compares the incoming data to the string "OFF"
  else if (strcmp(data,"OFF")==0) {
digitalWrite(led,HIGH);
  }

}
```

In the second code example, we are subscribing to the same channel that the first Particle Core board is publishing to. If the data string ON is sent, the second board will turn on the LED next to the USB port and if OFF is sent, the LED will turn off.

If you can't find anyone with a second Particle Core and you don't want to spend the extra money, don't worry, the Internet can be your friend. Next, I will show you how to connect and control your Particle Core from the Web.

HTML control

One of the cool things about the Particle Core is that you can implement something called **cloud** functions. The cloud part of course refers to data and services stored on the Internet, and with the Core boards you can include functions in your code that can be accessed by anyone with your device ID and token. Don't be alarmed to think anyone could hack into your board and take control; in order to do so, they would need your device ID and access token, which is a fairly long number and is impossible to guess.

This is not the same as the name you gave your board, which you can change. This name is just a reference name, which makes it easier for us humans to tell it apart from other boards. Your particular Core board has a unique identifier ID, which is hard-coded and cannot be changed. You can find your ID using the web IDE. If you press on the devices button (the one that looks like a crosshair), and press on the dropdown next to the name of your board, the ID will appear as shown in *Figure 8.7*:

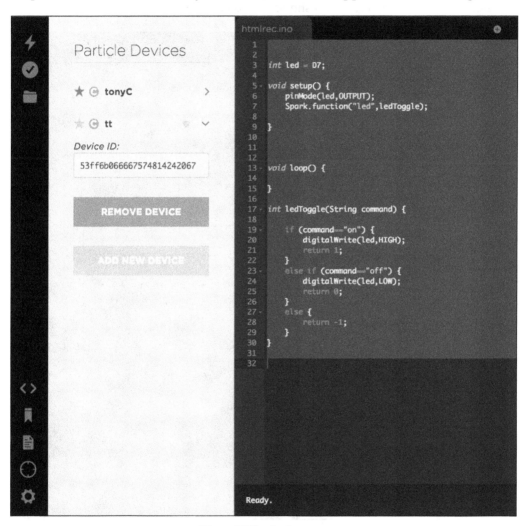

Figure 8.7 The device menu

We will need this ID later on. The second part, which makes it very hard to connect to your Core board if you don't have it, is the **Access Token**. A token is a piece of data used in network communication for different things such as login sessions, user identification, or to give privileges. It's like an access code that enables you to communicate with the Core board. You need the ID to find the board and you need an access token in order to connect to it. You can find the **Access Token** under the **Settings** in the web IDE. *Figure 8.8* shows the **Access Token**:

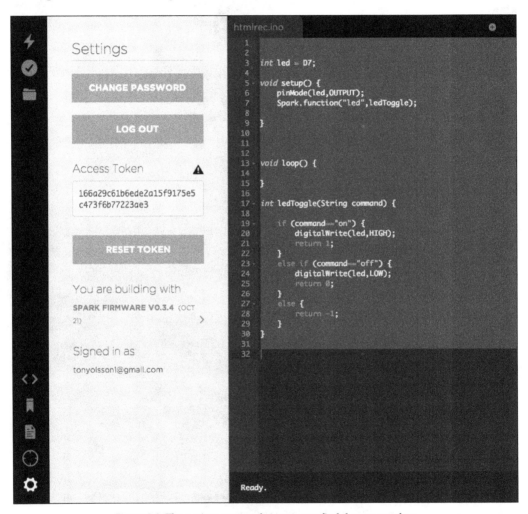

Figure 8.8: The settings menu where you can find the access token

The access token can be reset at any time for security reasons. Before you do so, make sure that you don't have something important connected to your Core board since you would need to update all devices with the new access token.

Now let's have a look at the cloud function. The next code example implements the function and triggers the internal function if the cloud function is called upon:

```
int led = D7;

void setup() {
    pinMode(led,OUTPUT);
    //Declare the function "led" and trigger the "ledOnOff" function
    Spark.function("led",ledOnOff);

}

void loop() {

}

int ledOnOff(String inData) {
//Check the incoming data a responded accordingly
    if (inData =="on") {
        digitalWrite(led,HIGH);
        return 1;
    }
    else if (inData =="off") {
        digitalWrite(led,LOW);
        return 0;
    }
    else {
        return -1;
    }
}
```

To use the function, you need to provide a **function key**, which is a string that is used to create a post request and a function name, which is the actual name of the function that is called. In the case of the code example just given, the function name is led. Note that the function name cannot be longer than 12 characters. If it is longer, the function will just skip any characters after the 12th character.

In the function, you can include yet another function to be triggered whenever the cloud function is called upon. In the case of the previous example, this is the ledOnOff function. Later, when we make a request to our Core board, we will do this by adding a string value. This value will be checked in the ledOnOff function and if the value is on, the on-board LED next to the USB port will turn on. If the value is off, then we will turn off the same LED. Once you have copied the code into the web IDE, flash it to your Core board.

Now let's make a simple website to try things out. If you are not familiar with HTML coding, don't worry because we will not go into it in any depth in this book. If you know HTML, this should get you started and if you don't I'm sure it will get you interested since the next part makes things very interesting, if you ask me:

```
<!DOCTYPE html>
<html>
<body>
Select and press send
<form action="https://api.particle.io/v1/devices/your-device-ID/
led?access_token=your-token" method="POST">
<input type="radio" name="args" value="on">Turn led on
<br>
<input type="radio" name="args" value="off">Turn led off
<br>
<input type="submit" value="Send to Core">
</form>
</body>
</html>
```

The website we have just created is by no means visually impressive, but is still very cool. It includes a simple form with two radio buttons and a normal button. From the radio buttons, you can select either on or off, and once you press **Send to core**, the website will make a request to the Core board and add the value of the button, which will activate the cloud function and turn the state of the on-board LED depending on the value.

In order for the website to connect to your Core board, you need to add your device ID and access token. Find your device ID and access token as shown previously in this chapter and change the HTML code where it says **your-device-id** and **your-token**. In order to make the website, copy the code into a text file and save it. Once saved, change the ending from .txt to .html and you are done.

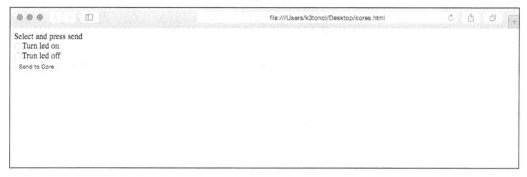

Figure 8.9: A simple HTML web page with control options

Open up the file in a web browser (double-click the file). *Figure 8.9* shows what the website should look like. Select the **Turn led on** option and press **Send to Core**. If everything works as it should, the on-board LED on the Core board should light up and your browser should display a message that looks like *Figure 8.10*:

```
{
  "id": "53ff6b0666675748142242067",
  "last_app": "",
  "connected": true,
  "return_value": 1
}
```

Figure 8.10 The response message from the Core board

The message displayed is information about the board and you can see the ID of your board, whether the board is connected to Wi-Fi, and last—the return value from the `ledOnOff` function. In *Figure 8.10*, the value is `1`, which means that the LED is on. If the value is `-1`, this means something went wrong. If so:

- Check all your connections
- Try re-flashing the Core board
- Check that you are connected to the Internet both on your computer and Core board

As you might have already figured out, having a web page connection to your wearable project opens up tons of possibilities for interesting projects, and in the last chapter we will get the chance to implement some of these cool functions. But we are not done yet; in the next part we will have a look at how to connect the Particle Core to online services such as IFTTT.

Connecting to IFTTT

If you are not excited about the possibilities of Particle Core yet, I am fairly sure a connection to IFTTT will get your creative juices flowing. IFTTT stands for **If This Then That** and is a web mashup service that brings together different online services and lets you connect them in different ways. Most big services with an open API are represented with IFTTT, where IFTTT lets you create what they call **recipes**. Recipes consist of **triggers** and **actions**. A trigger could be any device or service connected to IFTTT and the same goes for actions. For example, every time you upload a new picture to Instagram, you might want to store this photo in your Dropbox. With an IFTTT recipe, this process can be fully automated by connecting your Instagram account to your Dropbox account and letting IFTTT do its magic.

Now the cool thing is that with IFTTT, you can connect the Particle board to the same online services and automate different processes. For example, you can make a wearable button and every time you press it, it will store a note in Evernote about your geographical location or update your Facebook status. Possible services on IFTTT are called **channels**, which also include control options for Android phones. This means that you could activate a phone call or send an SMS from your Particle board. With the number of channels on IFTTT, the possibilities are basically endless. For the last project and chapter in this book, we will also use IFTTT as the backbone of the different functionalities in the project, but first let's start by getting familiar with IFTTT and trying out some basic interactions with the Particle board.

In order to get started, head over to: `https://ifttt.com/`, and sign up for an account, which is completely free. Once you are logged in, you can have a look at the different channels just to get a sense of what you can connect to what.

Before we get started with creating our first recipe, let's create some firmware for the Particle board so we can connect it to some services. We have two options, and we will have a look at these in this part. The first scenario is that we interact with the Particle Core and make something happen with IFTTT. The second scenario is that through interacting some service, we want something to happen on our Particle Core through IFTTT.

Monitoring data changes

Let's start with something simple by connecting a potentiometer to the Particle Core and if the value from the Particle Core reaches a certain point, we will make IFTTT post a Facebook status update for us. By using the `monitor` function for the Particle Core, we can open up variables for outside access and then we will use IFTTT to determine when to send the Facebook update. Start by flashing the following code to your Core board. If you run into problems, check that the Core board is connected to your Wi-Fi through the phone app and that you don't have any misspelling in your code:

```
int potVal = 0;

void setup()
{
   /* The variable function takes three parameters. The first is the
name of your variable. The name can be no longer than 12 characters.
The second one is your actual variable for the data you want to
monitor and the last parameter is the data type */
   Spark.variable("Pot value", &potVal, INT);
   pinMode(A0, INPUT);
}

void loop()
{
  potVal = analogRead(A0);
  delay(200);
}
```

The function called **P** enables the information to be accessed and called upon from the outside when connected to Wi-Fi from any device that has your Core ID and access token.

Next let's hook up a potentiometer to our Core board. To make things easy, we will hook everything up on the mini breadboard included in the Particle Core kit. *Figure 8.11* shows how to connect the wires in-between the Core and potentiometer:

Figure 8.11: Connecting a potentiometer to the Core board

There are many types of potentiometers out there and most of them will do for this test. The one I am using is a 10k potentiometer, but anything in between 1-10k will do fine. Potentiometers come in a variety of shapes and sizes and yours might not look exactly the same. Don't worry since most have the same pin layout. There are always three pins and the left and right pin is your power and ground respectively. The middle one is usually your signal pin but I can't account for every potentiometer out there, so if you are unsure please refer to the datasheet of your particular potentiometer. In *Figure 8.11*:

- The first wire, which the arrow on the potentiometer is pointing to, is connected to **GND**

- The middle pin connects to the analog pin **A1** on the Core

- The third pin is connected to **3V3**

Now we are ready on the hardware side so let's turn back to IFTTT. Navigate to the **My Recipes** tab and you should find something that looks like *Figure 8.12*:

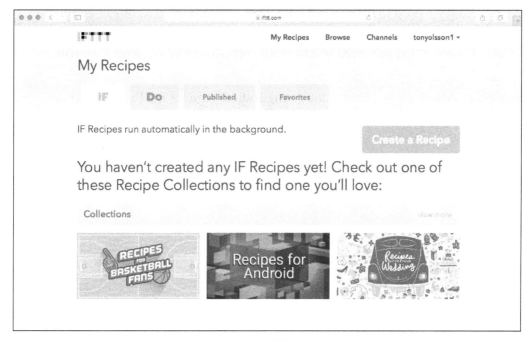

Figure 8.12: The IFTTT website

As you can see, there are two options for creating a new recipe. Either you can create an **IF** recipe or a **DO** recipe. In our case, we want to make an **IF** recipe since we want things to happen depending on our sensor data. Press the **Create a Recipe** button and follow these instructions:

1. When prompted with the trigger channel option, search for **Particle** and press the icon.

2. You will now be prompted with a login screen for your Particle account. Just log in using your Particle account information.

3. Now you will be asked to choose a trigger where you have the options of **New event publisher, Monitor a variable, Monitor a function result**, or **Monitor your device status**.

4. Choose the **Monitor a variable** option, but before you do, make sure your Core board is up and running with the code example.

5. Now you will be give the option to select a variable from your Core board. If you only have one Core board with the example code, the only thing shown in the drop-down menu will be your device and available variables, which in our case would be **Pot value**. Choose this variable.

6. Now we need to select a test operation, which will act as our `if` condition. Let's make it greater.

7. The last thing you need to add is the comparison value, which we will set to `500`.

Figure 8.13 shows the last part of the configuration page for the Particle channel:

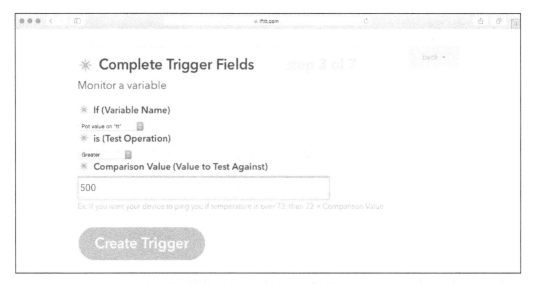

Figure 8.13: Creating the trigger for the recipe

As you might have figured out by now, you can set up the Core firmware with multiple monitor functions, which will appear in the variable list, and you can then choose your actions for IFTTT. In our recipe, we have now set it up so when the analog value we are monitoring on our Core board is greater than 500, we can make something happen. Now press **Create trigger** and we will move on to setting up our action.

You will now be prompted with the same option but this time for the intended action, which will take place once the analog value goes above 500. This time, peform the following steps:

1. Search for the Facebook channel and select it.

2. The first time you use a channel, you will need to fill in your account details for that service. Now, I'm just assuming most of you have a Facebook account, which might be presumptuous. If you don't have a Facebook account, you can choose one of the other services since most follow the same steps.

3. Once you are connected, your account will be displayed with the available functions, which are **Create a status message**, **Create a link post**, or **Upload a photo from URL**. Choose **Create a status message**.

4. Now type in the message you want to display on your Facebook page when the analog value goes above 500. As you can see in the message, there is already a message written, which says, **Variable is now Value** where variable and value have a grey background. This means that these are the actual variable name and value from your Core firmware. You can add your own message instead and still use these variables if you like.

5. Once you are satisfied with your message, press **Create action**.

Now you will be presented with a summary of your recipe, and if everything looks good, press the **Create recipe** button and you are done.

Log in into your Facebook account and open up your own personal page. Then turn the potentiometer at least 180 degrees to get above 500 and wait. You might need to refresh your browser if it does not do this automatically. After a few seconds, the message should appear in your status update. *Figure 8.14* shows what the standard message looked like on my Facebook:

Figure 8.14: Status update made from the potentiometer value

Don't forget to turn the potentiometer back since it will keep posting new messages as long as the value is greater than 500. Trust me, your Facebook friends will wonder what's going on if you don't.

We had a look at how to connect a simple sensor to the Core board and implement the monitor function in order to enable access to the variable data from the outside. We then used this data as a trigger for our IFTTT recipe where we connected it to a Facebook status update, which acts as our action in the recipe. Now let's have a look at how we can do the opposite, activating a function of the Particle Core from an outside web service.

DO – a function

For the DO function part we will make a mini SMS post box, which will give you a more physical indication that you have received a new SMS. The idea is that every time you receive a new SMS, we show a small light pattern on the Core board. This only works with Android phones so if you don't have one, you can choose one of the many other channels to work as your trigger. All of their functionalities are explained and the Core firmware in this part will remain the same. In order to be able to use functions on your Android device, you need to install the IFTTT app on your phone. You can find the app at the following link: `https://play.google.com/store/apps/details?id=com.ifttt.ifttt&hRL`.

Now let's start by connecting a few LEDs to the Core board. *Figure 8.15* shows how to connect four LEDs to the board using the mini breadboard:

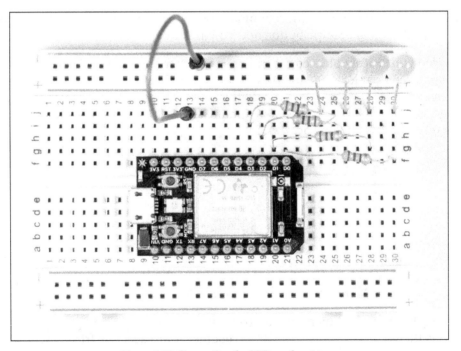

Figure 8.15: Connecting the LEDs and resistors

From the pins **D0-D3**, we have connected a 220Ω resistor for each of the pins. Each resistor connects to each of the positive legs of the LEDs. All the negative legs of the LEDs are connected to the ground ditch, which is connected back to Core board with a cable to **GND**.

In the firmware we will need to include a function, which will be accessible for IFTTT. This function will enable external control of the LEDs and it will display a simple light pattern. The patterns runs through all the LEDs forwards and then backwards:

```
//Declare all the digital pins into an array
int leds[]={0,1,2,3,4};

//Declare the function
int runPattern(String command);

//The led pattern
void runLed(){
//Run the forward/backward pattern 5 times
for(int t=0; t<5;t++){
for(int j=0; j<4; j++){
digitalWrite(leds[j],HIGH);
delay(200);
digitalWrite(leds[j],LOW);
delay(200);
}
for(int k=2; k>=0; k--){
 digitalWrite(leds[k],HIGH);
 delay(200);
  digitalWrite(leds[k],LOW);
 delay(200);
}
    }
}

void setup()
{
  // register the Particle function
  Spark.function("flahsLeds", runPattern);
  //Set all pins as outputs
  for(int i=0; i<4; i++){
   pinMode(leds[i],OUTPUT);
  }
//Run the pattern once just to test
runLed();
```

```
}


void loop()
{
//Do nothing in here
}

int runPattern(String command)
{
  //Look for the command "ledsOn"
  if(command == "ledsOn")
  {
//Activate the led pattern
 runLed();
 //Returns 1 if everything went ok
    return 1;
}
 //Returns -1 if something went wrong
  else return -1;
}
```

Now the firmware is set up so that every time the function flash LEDs is called from the outside, it will run the internal `runPattern` function if it is supplied with the right string, which is `ledsOn`. Flash the firmware to your Core board from the web IDE. Later, once everything is up and running, you can add more modes to your code by adding multiple activation strings and functions. But before I leave you to your own experimentation, let's first finish our IFTTT recipe and check that everything works.

Now head over to IFTTT and create a new `if` recipe:

1. For the trigger, choose **Android SMS**.
2. Then choose the **Any new SMS received** option.
3. Press the **Create trigger** button.
4. For the that action, choose the Particle channel.
5. Select the **Call a function** option.
6. From the drop-down list, choose the **flashLeds** function.
7. Add the sting **ledsOn** to the function input.
8. Press the **Create action** button.

Now you will be prompted with a summary of your recipe as before. If everything looks good according to the checklist, press the **Create recipe** button.

Make sure that your Particle Core board is flashed with the firmware and that the board is connected to your Wi-Fi. I guess most people only have one mobile phone so you will need to ask someone to send you an SMS. Once you receive it, you should see the LEDs start flashing. If you don't, always make sure you:

- First check your code for errors
- Check that the Particle Core is connected to your Wi-Fi
- Check that you have Internet access
- If these suggestions do not fix the problem, reset the Core board settings by holding down the **MODE** button until the RGB starts to flash and then reconnect the board using the Particle app

Summary

I hope you are as excited as I was about the possibilities the Particle Core board offered with Wi-Fi connectivity. To find out more about the functions of the Core board, I suggest you have a look Particles' online documentation: `http://docs.particle.io/core`.

In this chapter, we looked at how to set up and program for the Particle Core board. As you might have noticed, the process is a bit different from the normal Arduino board since this board is actually programmed over the Internet and not straight from the computer over USB. We also had a look at how to make a simple HTML web page in order to gain access to the Core board's functions. Last but not least, we had a look at how to connect the Core board to IFTTT, which offers endless mashup possibilities. We will progress on in the next chapter where we will be creating our very own smart watch.

As you can see, in this chapter there were a lot of steps in order to get the Particle board up and running, even though these steps were made very easy. In order to cover the basics, this chapter did not include full projects but acts a prequel to the next and final project.

Time to Get Smart

9

The time has come to end this book, but before it's all over, we will end with one final project that ties up all the knowledge and information in previous chapters. This chapter is a continuation of *Chapter 8, On the Wi-fly,* where we introduced the Particle Core board and how to interact with different online services. The same board will act as the hardware base in this chapter where we will create our own smart watch. But this project is based on my idea of the future, where all areas of even the smallest town will be covered by Wi-Fi and where smart watches, among other things, will connect straight to the Internet in a seamless manner. We might not be there yet but don't worry, the smart watch in this chapter will still work in the present. Let's not forget the best part about this project. Every time some one asks–what's that on your arm? You can reply, it's a smart watch I built myself.

I have always had a problem with smart watches. First of all, I don't like that all watches you could buy up to the point I am writing this book are dependent on a mobile phone. The second part I don't like is that these devices never seem to do exactly what I want them to do.

The solution, then, is to build your own, and with off-the-shelf components, you can come very close to the real thing. Or in some ways, better than the real thing since with this watch you can customize it in any way you see fit.

By the end of this chapter, you should have a good grasp on how to build your own wearable projects using the tools and knowledge in this book. As I have said before, the knowledge in this book is an introduction, but the goal is for you to build upon this knowledge in order to truly make the project your own. Make modifications, tweak the designs, and make any changes you see fit in order to make the knowledge yours. But before we end, let's make a watch.

Components

For this project, you will need the components mentioned in the following list. All the components listed can be swapped for similar components, and in *Figure 9.1*, you will find the components I used. If you do swap the components, make sure that you keep track of the dimensions of your components. The design for the watch in this book is based on the components shown in *Figure 9.1* and leaves very little room for modification since we are trying to achieve as small a casing as possible.

- A Particle Core board
- A 3.7V lithium battery
- An Adafruit 128 inch x 64 inch OLED screen
- An Adafruit Battery charging circuit
- A 1 mm plastic sheet material
- Leather or heavy duty material, at least 30 cm x 10 cm
- A soldering iron
- An Exacto knife
- Leather sewing awls and heavy duty thread
- Eyelets and an eyelet tool
- A mini breadboard

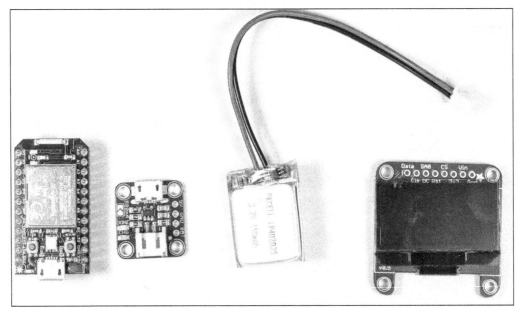

Figure 9.1: The components needed for the smart watch

If you want to pack the full punch with your watch, you could consider including a GPS, however this will make the design much bulkier. The design for this chapter aims for as thin a design as possible. However, if you want, you can add many of the sensors used in prior chapters of this book and attach them to the watch later. In this chapter, we will stick to the basics of making a smart watch and limit ourselves to the use of online services. But don't worry, there is still a lot that can been done with this watch.

In this project, we will also include a battery charging circuit so that we can charge our watch without opening up the project. The idea is that later on, when finishing the project, we will glue everything together in order to make the project somewhat weather-proof. We will only leave a small hole for the USB micro connector of the battery charger in order for the watch to be recharged.

Let's get started

We could start building the watch straight away since the Particle Core board can be programmed wirelessly and there is no need to have the parts exposed. However, when it comes to any form of prototyping, this is never a good idea, as a lot of things can go wrong on the way, so it's better to break down any project into parts, as we have done throughout this book.

The star of the show will be the screen we use for this project. There is a large variety of different small screens that can be used, and I have chosen the SSD1306, 1.3 inch, 128 x 64 OLED screen from Adafruit. For this project, we will be using the same screen libraries from Adafruit as in the previous chapters, and they have a large selection of screens that work with this library, so you can choose any one you like. The screen I have chosen is black and white, but you can switch it to any of the color screens as well. The design for this project was inspired by a mix of 80s Casio watches and leather straps, so I went for the black and white screen as I think this gives it a nice retro look. If you do change the screen, make sure that you modify the design of the wristband for your watch so the screen fits the design.

Let's start by connecting the screen to the Core board to see whether we can display some graphics. Most of these screens come without the pinheads soldered, which is perfect since we want to use cables instead. For this project, I recommend using soft cables for all the connections as we need to keep things flexible. We will not solder the screen to the Core board yet we want to try out all our components first.
In the meanwhile, we will use a mini breadboard.

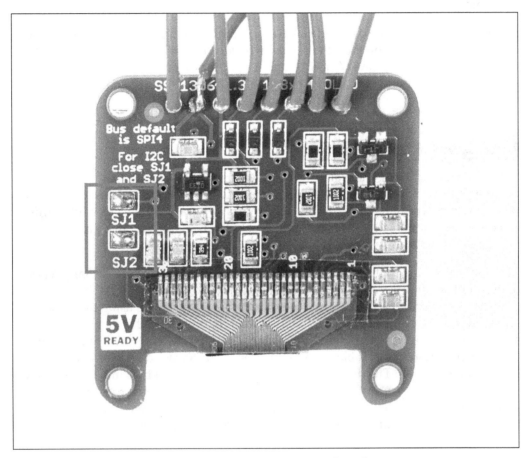

Figure 9.2: Soldering the SJ1 and SJ2 pad together

With the 128 x 64 display used in this chapter, you will need to modify the PCB in order to get it to work over I2C. The reason for using I2C is that this will make it easier to interact with both the screen and the sensor accelerometer/gyro/compass sensor at the same time. *Figure 9.2* shows the pads named SJ1 and SJ2 on the back of the screen pcb that need to be soldered together to tell the board to communicate over I2C.

Once the pads are soldered together, do the following:

1. First, cut and strip 6 cables to approximately 10 cm in length.

2. Solder the cables to the OLED screen on by one, but skip the VIN since this is not used.

3. Once done soldering, connect all the wires to the breadboard and Core board in the following sequence:

   - **Data** connects to **D0**
   - **CLK** connects to **D1**
   - **RST** connects to **D4**
   - **SA0** connects to **GND**
   - **3V3** connects to **3V3**
   - **GND** connects to **GND**

Now that we have everything connected, let's turn to the web IDE and get started with the code. Many libraries are available from the web IDE and can be installed with a click of a button, but these libraries rely on them being available on GitHub. In the next part, I will show you how to add your own libraries to the web IDE in case you run into trouble or need to modify an existing library. We will still use the Adafruit libraries, but we will manually add them to your firmware this time.

1. First, start by opening up a new sketch from the web IDE.

2. Next, add a new tab by pressing the **+** sign at the top right corner, as shown in *Figure 9.3*, which will open up a new tab for you.

3. Name this tab **Adafruit_SSD1306.h** and this will automatically generate an **Adafruit_SSD1306.cpp** tab for you as well.

4. Next, head over to the GitHub repositories: `https://github.com/tchoney/Adafruit_SSD1306`.

   Copy the code from `Adafruit_SSD1306.h` into the tab of the web IDE with the same name. I have forked this repository from Paul Kourany, who has made a port of the original Adafruit library to the Particle Core board. Forking a library means making a copy of it so you can modify the code.

5. Do the same for the `Adafruit_SSD1306.cpp` code.

6. Now we have to do the same operation for the `Adafruit_GFX` library. Press the **+** sign in the IDE and name the .h tab as `Adafruit_GFX` library and the `.ccp` tab will automatically generated.

7. Copy the code from the GitHub repository mentioned earlier to the tabs with the corresponding names.

Figure 9.3: Manually adding libraries to the web IDE

Now, we have successfully included the libraries and we are free to make modifications to them if needed. The library is set up for 128 x 64 display in the `Adafruit_SSD1306.h` file. For example, if you are using the 128 x 32 pixel version, you would need to modify the `Adafruit_SSD1306.h` file, which includes comment on how to do so. In order to test that everything is working, add the following code to the `.ino` tab, which is the main tab for your firmware. When done, flash the firmware to the Core board.

```
#include "Adafruit_GFX.h"
#include "Adafruit_SSD1306.h"

#define OLED_RESET D4
Adafruit_SSD1306 display(OLED_RESET);

#define NUMFLAKES 10
#define XPOS 0
#define YPOS 1
#define DELTAY 2

#define LOGO16_GLCD_HEIGHT 16
#define LOGO16_GLCD_WIDTH  16

#if (SSD1306_LCDHEIGHT != 64)
#error("Height incorrect, please fix Adafruit_SSD1306.h!");
#endif

void setup()   {
  Serial.begin(9600);

  // Init display
  display.begin(SSD1306_SWITCHCAPVCC, 0x3C);  // initialize with the
I2C addr 0x3D (for the 128x64)
  //Clear the memory
  display.clearDisplay();
}
```

```
void loop() {
  display.clearDisplay();
  display.setTextSize(2);
  display.setTextColor(WHITE);
  display.setCursor(0,0);
  display.println("Time to");
  display.display();
  delay(2000);
  display.clearDisplay();
  display.setTextSize(2);
  display.setTextColor(WHITE);
  display.setCursor(0,0);
  display.println("Get smart!");
  display.display();
  delay(2000);

}
```

If the message from the code does not appear on the screen:

- Check your cable connections
- Check that the code is correct
- Check that your Core board is connected to your Wi-Fi
- Try re-flashing the board

Watch design and soldering

Now it's time to move on to the design of the watch; as mentioned before, you can incorporate any kind of sensors you like into the design of your watch, but take into account the size of any component you add. In order to turn your components into a watch, we will base the design on the template shown in *Figure 9.3*. Ideally use this template to cut the shape out of a 1 mm thick plastic material. This can be any material as long as it's sturdy enough but has some flexibility to it. The plastic material I used, also shown in *Figure 9.4*, is a fairly common 1 mm plastic material found in arts and crafts stores.

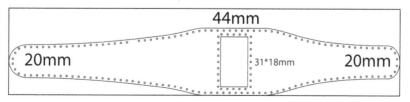

Figure 9.4: The template design for the watch strap

The inspiration for the design comes from traditional wristwatch leather crafts where they often use plastic inlays to make the straps more ridgid. The leather inspiration does not stop there, since we will be using leather as our other casing for the watch, giving the design a nice mix between high and low tech.

As you can see in *Figure 9.4*, the edges are covered with small holes. The idea is to use these holes later on as guidance once we sew our watch together. You can cut the template by hand since 1 mm plastic is fairly easy to cut with a pair of normal scissors. But, once again, if you do have access to a laser cutter, it would be the preferred choice since the spacing between the holes can be hard to achieve by hand. There are specialized watch-making tools for making holes like this, but if this is the only watch you plan to make, a nail and a hammer will do just fine.

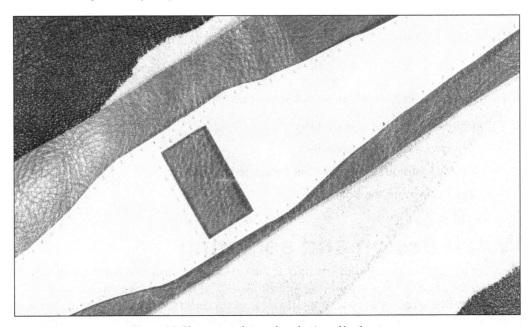

Figure 9.5: The cut template and a selection of leather types

In *Figure 9.5*, you will find the cut template and a selection of different leather types. Any type of leather you prefer should work for this project, but my recommendation is to choose one that is not too thick. I would stay below 3 mm or else it might be hard to sew when it's time to stitch things together. There are also synthetic leathers available if you are not to keen on the idea of using real leather, or you could use fabric as well. Don't forget that this might not make the watch as weather-resistant as with leather. In any case, this watch will never be waterproof so don't go swimming with it.

Before we start soldering the components together, it is a good idea to place them onto the template in order to figure out the best location. In *Figure 9.6,* you will find the layout I was aiming for, which gives you a good general direction. In my case, I have very thin wrists so it is a good idea to put the components in place using sticky tape in order to wrap the template with the components around your wrist to see whether the placements works for your wrists. The important parts are that the screen needs to line up with the hole in the template and the micro USB connector on the battery charger. The USB connector needs to be placed close to the edge so that we can access it later on once everything is stitched together.

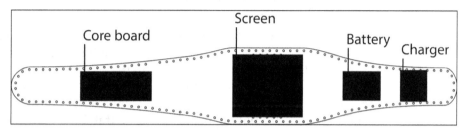

Figure 9.6: The placement of components on the watch template

Once you feel your components are located in the right place, you can mark their position onto the template using a pen. Before we attach and solder them together, we need to make some modifications to the Core board and the USB battery charger. As you might have noticed, the JST connecter and the pins of the Core board are not optimal for a flat design like in this project, and we can fix this by desoldering the pins and JST connector since there is no need for them.

Desoldering

There are many ways you can desolder components and it is truly an art in itself. If you are new to desoldering electronics, I would recommend starting with cheaper components. An old broken radio is a perfect training ground as it includes many types of components in different sizes and shapes. There are special materials available for desoldering components, such as solder wicker and suction pens, but personally, I found that the best way to do it is to use a soldering iron and a pair of tweezers to grab and pull the components. How to desolder electronics is hard to illustrate in a book so I recommend having a look at these videos to get a sense of what we are dealing with:

https://www.youtube.com/watch?v=77JgIqraX_I

https://www.youtube.com/watch?v=Z38WsZFmq8E

There is no single way that is the best way, so I would watch as many videos you can in order to get a good grasp on the concept. Then, start training on your own on any kind of electronic junk you might have lying around. However, two ways you could do this is:

- To heat the solder, hold a component in place and then drop it hard on to a table. When the component hits the table with force, if the solder is hot, the solder will usually fall off on to the table. Be very careful using this technique as you do not want to splash hot solder onto your finger or any flammable materials of surface. Never use a table that is not meant for working on, since the hot solder can leave a mark.

- The alternative approach is to heat the solder pads and pull the component using tweezers or a pair of needle nose pliers. Again, you heat the solder holding the components in place, and then you pull the components using the tweezers and your other hand. Make sure that you never heat a component for too long since this can damage the components. Only heat it for a few seconds and then try pulling the component. If it does not come off, let it cool down for a few seconds before you try again. Beware of heating the tweezers as they can get hot fast.

Once you feel confident in desoldering, you can get going with removing the JST connector from the battery charger assuming you are using a similar charger and the male pin headers from the Particle Core board. If everything goes to plan, you should end up with something that looks like *Figure 9.7*.

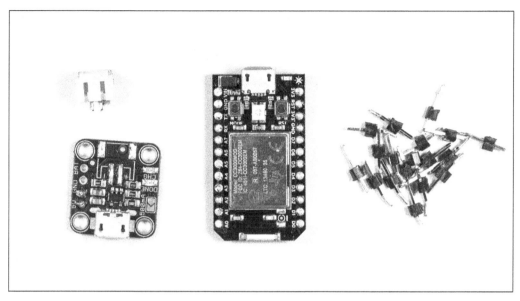

Figure 9.7: The JST connector and male pins desoldered

In order to remove all the male pin headers from the Core board, the trick is to cut the plastic that holds all the pins together using a pair of mini side cutters, or similar, and then you can desolder the pins one by one instead. In order to remove the JST connector from the USB charger, I recommend starting the ground pads on the side. Start with one side heating the solder and pull it up as much as you can, and then do the same for the other side. Do this a couple of times and eventually the connector clears from the pcb board. Then, heat the positive and negative connectors at the same time and simply remove the JST connector.

After you have finished desoldering, have a look at your pcbs and check that you haven't soldered connections together by accident. Take your time desoldering the components in order to not damage any of the other components. For future investigations into desoldering, I would recommend looking up solder suckers or the desoldering braid shown in the video link mentioned earlier. Personally, I don't prefer these tools, but this should not stop you from trying them out. As I said before, when it comes to soldering and desoldering, there are a lot of different techniques and tools, and you should investigate different approaches and try to find a work method that fits your needs.

Connecting the pieces

Now it is time to solder all of the components together. First, use your template and the marks of the components to measure out the cables needed to connect everything. The screen already has the cable connected to it so you don't need to make new ones, but you will need two cables for power and ground connected from the USB charger to the Core board.

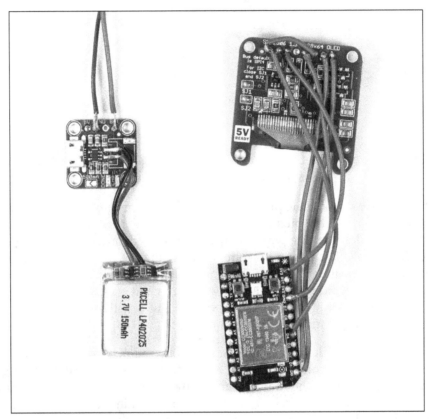

Figure 9.8: The battery and charging circuit soldered together with the screen and Core board

Figure 9.8 shows the cable connections that need to be made at this stage. Right now we will not connect the battery to the Core board as this would leave the Core board on, which we don't want while working on the watch. As you might suspect, this watch will be on at all times. This was a conscious design choice since adding an on-off button would complicate things. In order to keep the design as sleek as possible, I chose not to include one since it is not really needed. Even with a small battery, the watch will last a long time depending on the firmware you put on the Core board. If the battery does run out, you will simply need to connect a USB cable and it will turn on again and charge the battery.

Now, back to the connection that needs to be made. For the battery charger, you will need to do the following:

1. Connect a cable to the **GND** and the **BAT** pin of the charger PCB and leave the other ends unsoldered for now. Later, we will solder the ends to the **VIN** and **GND** on the Core board.

2. Cut the male JST connector off the wires on your lithium battery. Don't cut both wires at once since this will short-circuit the battery.

3. Start by soldering the red cable to the solder pad marked **+** where you have removed the female JST connector from the battery charge PCB.

4. Then, solder the black cable to the pad marked with a **–** sign.

Once the battery is in place, try connecting the USB cable to the battery charger and the other end to a computer. This is done in order to check that the soldering is good, and if it is, the battery should start to charge. If you are using the same USB battery charger shown in *Figure 9.7*, the LEDs should start glowing red if the battery is uncharged. Once it is charged, the LED should turn green.

The connections that need to be made from the OLED screen to the Core board are the same as when you were testing the screen on your breadboard:

- **Data** connects to **D0**
- **CLK** connects to **D1**
- **RST** connects to **D4**
- **SA0** connects to **GND**
- **3V3** connects to **3V3**
- **GND** connects to **GND**

As you can see in *Figure 9.8*, the **SA0** pin is bridged to the **GND** pin on the screen and then connected from the same pin to the **GND** on the Core board. Once you have the connections soldered, it is time to prepare one side of the watch template with some leather.

Leather time

Once you have soldered your components together, it is time to bring out the needle and thread as we will need to stitch the last part of the construction. There are a few steps to this next part in order to finish the watch. I recommend that you first read through this part to get a good overview of the steps so you can make the necessary preparations and gather your tools.

You will also need the piece of leather you have selected. *Figure 9.9* shows how the watch template is placed onto the piece of leather I used for the watch in this chapter. There is no need to cut the leather yet, since once we will start stitching, you will note that the components on top of the template will add to the demotions of the watch and having some extra length to the leather really helps. The amount in *Figure 9.9* is a bit overregulated, but I would leave a minimum of 1 cm around the edge of the template if you are cutting yours from a bigger piece of leather.

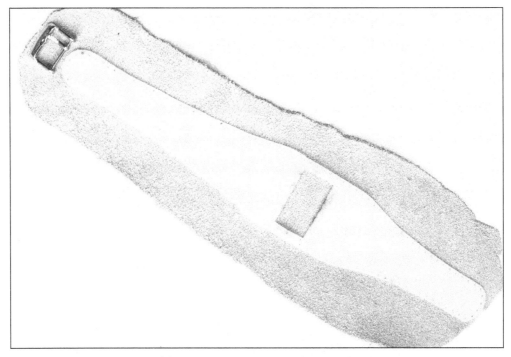

Figure 9.9: Attaching the template to the piece of leather

The next step is to cut a hole for the screen. *Figure 9.10* shows how the screen was marked out on the inside of the leather and then cut from corner to corner using an exacto knife.

 Make sure you mark the inside and not the outside of the leather so you don't end up with markings on the side that will make your outer casing.

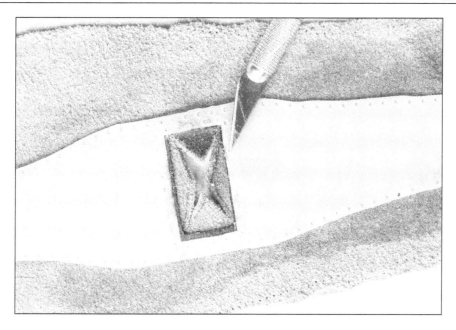

Figure 9.10 Cutting a hole for the OLED screen

1. Once the hole is made, we will attach the watch template and leather together.

2. Cutting the hole into the leather should leave you with four flaps as in *Figure 9.10*. Folds these flaps one by one so that they end up on the inside of the template to the left, as shown in *Figure 9.11*.

3. Use super glue in order to fix them in place. Make sure you don't use too much super glue since it has a tendency to make leather very hard.

4. Once you have folded and glued the flaps in place, you can try lining the hole up with your OLED screen.

If you are happy with the result, we can move on. Remember, what you are looking for is for the hole to cover enough of the edge of the screen so you can't see the edge. If you are scared that you have covered too much of the edge, don't worry. Later on, we can change the position on anything we display from our code, so everything lines up nicely on the screen. Once the leather is fixed to the template, you can cut the flaps so they don't extend beyond the side of the watch. Then, start stitching around the screen using the template holes. This is to secure the template in place and to give a nice border around the screen.

You have countless options here in terms of color design. Thread and leather both come in any color you could imagine, so there are a lot of options. Note that you want to use a heavy-duty thread for this project. The thread needs to be strong in order to hold everything together. If you are not sure on which thread to use, I would recommend you head over to your local sewing shop and ask around.

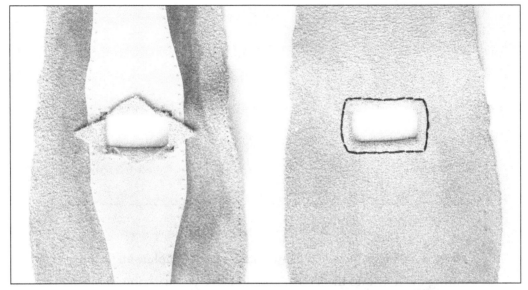

Figure 9.11: The flaps glued on the inside and the stitching around the screen from the outside

Once you are done with stitching your screen border, it's time to start adding your components to the template.

1. Start by lining up your screen to the hole in the leather, and once you have it lined up, secure it in place using hot glue around the edges of the screen.

2. Follow the component layout in *Figure 9.6*, then glue the Particle Core board, battery, and battery charger in place.

3. Make sure the micro USB port on the battery charger lines up with the edge of the template since you will want to be able to access it once on both sides of the watch.

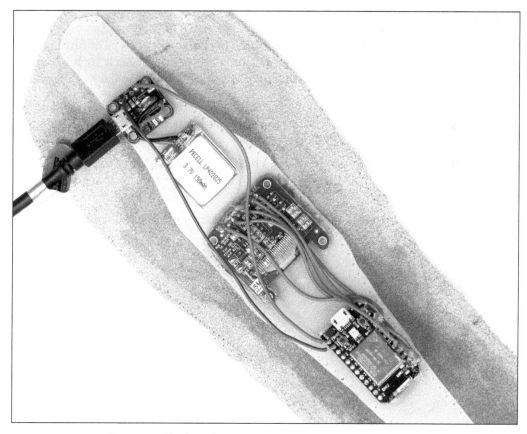

Figure 9.12: Checking the components once placed on to the template

When you glue the components, remember any modifications to the design you made depending on your wrist size. I also recommend that you hot glue your entire solder joint to add some extra sturdiness to your circuit. Put some glue on the solder pads and all the way up to the cable casing in order to secure things. Once all the components are glued in place, it's time to connect the battery charger to your Core board. Start by soldering the 5V connector from the battery charger to the **VIN** on the Core board. When the last cable is connected, the Core board should light up. From now on, your watch will be on, so be careful not to poke your circuit with any metal as this might cause a short circuit.

Finishing up

Now that you have your circuit in place, it is time to take the final steps of turning everything into a watch. The first thing we need to do is to add a watch latch to the strap of the watch. This is done on the second piece of leather that will cover the watch.

1. In *Figure 9.13*, you can see how this can be done by cutting a 2 x 2 cm flap from the leather. Make sure that this piece of leather will fit the watch plus the latch flap.

2. In order to hold the latch in place, a hole is cut at the base of the flap, and then the flap is pulled through the latch and glued in place, as seen on the left-side of *Figure 9.11*.

3. The pin in the middle of the latch should go through the hole in the leather.

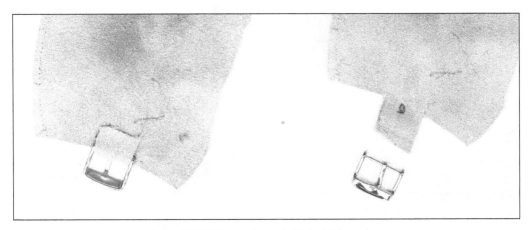

Figure 9.13: Cutting a flap to hold the latch in place

When the latch is in place, it's time to bring out the needle and thread. You want the latch flap to extend the short strap of the watch template, and *Figure 9.12* shows how the second piece of leather covers everything.

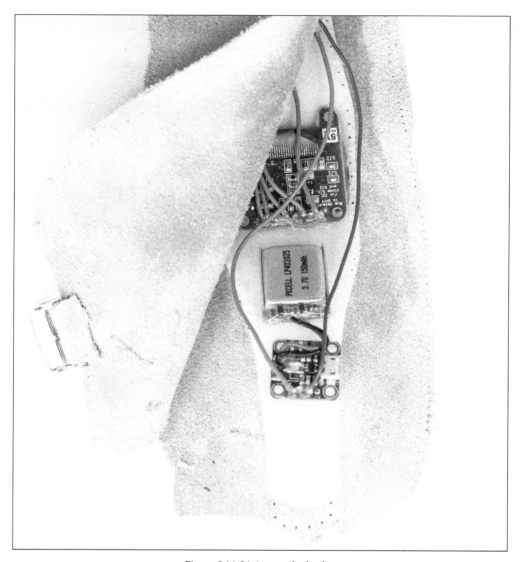

Figure 9.14: Lining up the leather

When it comes to stitching everything together, use the holes in the template to guide you. Note that you will need to give some slack to one of the sides of leather since the components add some dimensions to your watch. When it comes to stitching techniques, you should go with the one you feel most comfortable with, however, I found the best way to get good results is to go over and under all the way around the watch, as in *Figure 9.15*.

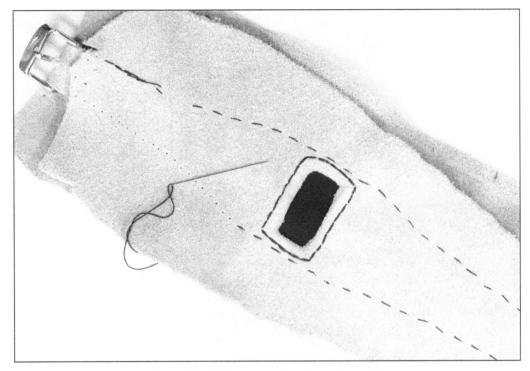

Figure 9.15: Stitching around the edges of the watch

Once you have reached the end of one side, reverse the stitching and go the other way around to fill in the gaps.

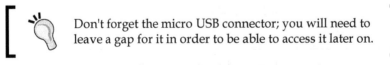

Don't forget the micro USB connector; you will need to leave a gap for it in order to be able to access it later on.

You might want to put on a pot of coffee for this stitching part since it will take some time, but keep the result in mind to stay motivated. Once the stitching is done, you can cut the watch into shape using a good pair of scissors or your exacto knife. Remember not to cut too close to the template and the stitching. I would suggest you leave about 2-3 mm on the edge of the watch template.

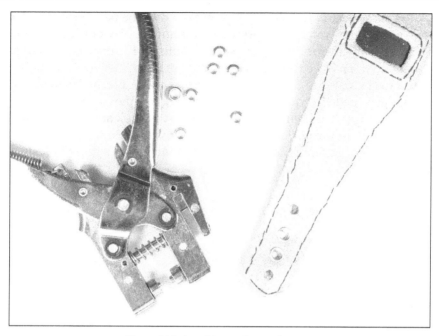

Figure 9.16: Attaching eyelets to the strap

Once you are done stitching, you should end up with something that looks like the watch shown in *Figure 9.16*. The last step is to punch some eyelets into the longer strap on the watch. You can buy an eyelets kit fairly cheap, where you get some eyelets and a punishing tool. These tools are fairly self-explanatory, but make sure you check and feel the strap so there are no hidden wires in the area where you plan to punch a hole. The number of eyelets all depends on your wrist size and style preferences.

Once you have reached this point, I hope you fingers are not too sore from stitching leather because it is time to do some coding for the watch.

A smorgasbord of functionality

As I mentioned before, the version of smart watch we are building in this chapter is based mainly on receiving notifications. The idea is to base most of the notifications on IFTTT since it has a nice infrastructure to connect different services and saves us a lot of code and time. We could connect to most of the services available through IFTTT without using IFTTT and using the open APIs available from the different services. However, IFTTT makes it much easier to connect different services without the need to learn about all the APIs from each individual service. But remember that your watch project does not need to end with this chapter since what I present here is just the tip of an iceberg of possibilities.

Now back to the code part, the following example shows the basic principle that we will use to connect your watch to different services and display notifications on the screen. The first firmware for the watch uses the same IFTTT function as presented in *Chapter 8, On the Wi-fly*. Once the following firmware is updated to the Core board, you will need to head over to the IFTTT website and create a recipe from your account where any updates made to your Facebook status triggers the "getMail" from your Particle channel. If you don't remember the steps, have a look at *Chapter 8, On the Wi-fly* again to refresh your memory. For the following example to work, you need to include the Adafruit_GFX and the Adafruit_SSD1306 libraries as previously used in this chapter:

```
#include "Adafruit_GFX.h"
#include "Adafruit_SSD1306.h"

#define OLED_RESET D4
Adafruit_SSD1306 display(OLED_RESET);

#define NUMFLAKES 10
#define XPOS 0
#define YPOS 1
#define DELTAY 2

#define LOGO16_GLCD_HEIGHT 16
#define LOGO16_GLCD_WIDTH  16

#if (SSD1306_LCDHEIGHT != 64)
#error("Height incorrect, please fix Adafruit_SSD1306.h!");
#endif
//Declare the function
int fbStatus (String topic);

void setup()    {
    Serial.begin(9600);
    //Initialize the function
    Spark.function("getFbStatus", fbStatus);
    // Initialize display
    display.begin(SSD1306_SWITCHCAPVCC, 0x3C);
//Clear the memory
    display.clearDisplay();
    display.setTextSize(2);
    display.setTextColor(WHITE);
    display.setCursor(10,11);
    display.println("ON");
    display.display();
    delay(2000);
```

```
}


void loop() {
  //Leave the loop empty
}

int fbStatus(String topic){
    if(topic !=""){
    display.clearDisplay();
    display.setTextSize(2);
    display.setTextColor(WHITE);
    display.setCursor(10,11);
    display.println(topic);
    display.display();
    delay(2000);
    }
    else return -1;


}
```

In the code example, once the function is triggered it will display any message included in the trigger where I have included the actual message posted from Facebook into my trigger. Of course, you can change this to whatever you want. If you noticed, I have shifted the starting point of the cursor in the example just given. The reason for this is that on the watch presented in all of the figures in this chapter, the leather rim of the hole for the screen covers the screen slightly. This was intentional, as I did not want the edge of the screen to be visible since we can move the cursor so that the text lines up with the leather instead. If the rim of your screen is slightly covered, you can play around with the cursor configuration until you are satisfied. The cursor is calculated from the left top corner in X and Y, where increasing X would move the cursor to the right, and increasing Y would move the cursor down.

For the next and final code example for the watch, I have added a few more notifications just to show off some of the many possible functions. How about getting a notification when your favorite person on Instagram posts a new picture, or how about getting updates on any price changes on a particular product you are looking to buy? Of course, this would not be a watch if it did not tell the time, so I have included some code that displays the time and date, which the watch will receive from the Particle servers. This means that you know the time received is reliable because it relies on the same atom clocks on the GPS satellites that we connected to in a prior chapter.

Right now, this example is set up for testing purposes and to get you going with your watch. The time and date is always shown on the screen until an event is triggered from IFFTTT. Then, the information corresponding to the particular IFTTT recipe is shown for 2 seconds before turning back to showing the watch and date:

```
#include "Adafruit_GFX.h"
#include "Adafruit_SSD1306.h"

#define OLED_RESET D4
Adafruit_SSD1306 display(OLED_RESET);

#define NUMFLAKES 10
#define XPOS 0
#define YPOS 1
#define DELTAY 2

#define LOGO16_GLCD_HEIGHT 16
#define LOGO16_GLCD_WIDTH  16

#if (SSD1306_LCDHEIGHT != 64)
#error("Height incorrect, please fix Adafruit_SSD1306.h!");
#endif
//Declare the functions
int fbStatus(String topic);
int displayTime();
int instaGram(String insta);
int bestBuy(String pebble);
int missedCall(String pNbr);

void setup()    {
    Serial.begin(9600);
    //Initialize the functions
    Spark.function("getfbStatus", fbStatus);
    Spark.function("getBSinsta", instaGram);
    Spark.function("getBestBuy", bestBuy);
    Spark.function("getMissCall", missedCall);
  // Init display
    display.begin(SSD1306_SWITCHCAPVCC, 0x3C);  // initialize with the
I2C addr 0x3D (for the 128x64)
  //Clear the memory

}

void loop() {
```

```
//Show the time and date
displayTime();


}
//If your Facebook status is updated show the status
int fbStatus(String topic){
    if(topic !=""){
    display.clearDisplay();
    display.setTextSize(2);
    display.setTextColor(WHITE);
    display.setCursor(10,11);
    display.println(topic);
    display.display();
    delay(2000);
    }
    else return -1;


}
//If Brittany Spears updates her Instagram let me know
int instaGram(String insta){
    if(insta !=""){
    display.clearDisplay();
    display.setTextSize(2);
    display.setTextColor(WHITE);
    display.setCursor(10,11);
    display.println("New Briteny");
    display.display();
    delay(2000);
    }
    else return -1;
}
//If the price on the latest Pebble smart watch changes let me know
int bestBuy(String pebble){
    if(pebble !=""){
    display.clearDisplay();
    display.setTextSize(2);
    display.setTextColor(WHITE);
    display.setCursor(10,11);
    display.println("Pebble price");
    display.println(pebble);
    display.display();
    delay(2000);
    }
    else return -1;
```

```
}
//If I have a missed phone call on my Android device show the number
int missedCall(String pNbr){
    if(pNbr !=""){
    display.clearDisplay();
    display.setTextSize(2);
    display.setTextColor(WHITE);
    display.setCursor(10,11);
    display.println("Missed call");
    display.println(pNbr);
    display.display();
    delay(2000);
    }
    else return -1;
}
int displayTime(){
    /*Set the courser and display the time in hours, minutes and
seconds. Remember this is standard time so you need to add or decrease
the hours depending on your time zone*/
    display.clearDisplay();
    display.setTextSize(2);
    display.setTextColor(WHITE);
    display.setCursor(15,15);
    display.print(Time.hour());
    display.print(":");
    display.print(Time.minute());
    display.print(":");
    display.print(Time.second());
        /*Set the cursor and display the date in days, months and
year*/
     display.setCursor(15,40);
     display.print(Time.day());
    display.print("/");
    display.print(Time.month());
    display.print("/");
    display.println(Time.year());
    display.display();
    delay(1000);
}
```

If you have problems uploading the firmware to the Core board, make sure you check the following:

- That your board is powered on.
- That your Core board is connected to the same Wi-Fi network.
- If nothing works, try resetting the board and holding down the mode button. Once inside the watch, you will not see the mode button, but you should be able to feel it from the outside. Hold it down for 5 seconds, and then try to reconnect your board to your Wi-Fi using the Particle mobile app.

This is what your final watch should look like:

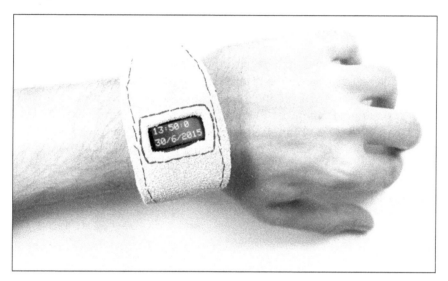

Figure 9.17: The final watch result

The end of the beginning

We have finally reached the end of the chapter, and so the end of the book. In this chapter, you learned a lot more about the Core board and how to add libraries manually to the web IDE. You also learned more about how to interface with OLED screens over both SPI and I2C and how to connect everything in a small form factor. This chapter also introduced you to some leather crafting and how to shape your project into a watch.

On the programming side, we extended the introduction made in *Chapter 8,*
On the Wi-fly to program the Core board. While we have created a working Wi-Fi
connected watch with some notifications from different online services, we have
barely scratched the surface of the possibilities for your smart watch. Hopefully,
you have gained enough insight to start developing the project in the future in
order to meet your needs and wishes.

In fact, this to me is what wearables are all about, bending technology to the will
of the user. I hope I have shown that you do not need to wait for manufacturers to
eventually make something that fits your preferences. The technology already exists,
and with a bit of knowledge, you can create your own devices. As with any craft,
the more time you spend doing it, the better you will become at it.

As I mentioned at the beginning of the chapter, there aren't that many cities around
the world that have full Wi-Fi coverage. This might be a problem for those readers
who want to use their smart watch beyond the coverage of their own Wi-Fi router.
In the meanwhile, you can share the Internet using most modern phones as they
can act as a Wi-Fi hotspot. Just set one up on your phone, connect to your watch,
and you are good to go. Remember that this watch is not dependent on any particular
operating system, so it does not matter whether you are running Android, OS X, or
Windows. Another potential problem is switching between Wi-Fi networks. In order
to do so, you will need to use the Particle app on your iOS or Android device.

You can also use a computer to set up your Particle Core board over USB. On the
following website, you can find more information on how to do so:

```
http://docs.particle.io/core/connect/
```

Before we end this chapter, I would like to suggest that you have a look at the
following recourse in order to progress your own smart watch. First, there is a lot
of functionality in the Particle Core board that we did not have room to cover in
this book. On the following site, you will find more information on the Core board,
which might be useful for your project:

```
http://docs.particle.io/core/start/
```

IFTTT still has a lot to offer in terms of notifications that can be made and just
browsing through the different channels will certainly inspire new possibilities.

```
https://ifttt.com/
```

The watch presented in this chapter is the bare bones of what I consider to be a smart watch for you to improve on. Even if there is a lot of functionality you can add on the software side, modifying the hardware would also create some new possibilities. Many of the sensors used in this book were picked for their versatility and some could be added to this project. A natural step would be to add an accelerometer and the possibilities will multiply yet again, or maybe think of ways you can connect some of the projects to one and other. My recommendation is to not stop here, but to keep on developing all of the projects in the book.

Unfortunately, this book has to end at some point and this is it. I hope you have enjoyed the projects presented, and as always,

Happy prototyping!

Index

R

radio frequency identification (RFID) 97
recipes
  about 150
  actions 150
  triggers 150
Red Bear GitHub
  URL 115
RedBearLab
  URL, for Android 131
  URL, for IOS 131
RX pin 82

S

SCL pins 78
SDA pins 78
sensors
  about 20
  bend sensor 20-25
  pressure sensor 25-29
Serial monitor
  about 6
  blinking 14, 15
  external LED 14, 15
  speed blinking 16, 17
Serial.print()command 25
Serial.println() command 25
servomotor, NFC
  connecting 103, 104
software
  installing 3, 4
  using 3, 4

T

TSL2561
  about 40-44
  URL 41
TX pin 82

U

Uno Arduino boards
  about 8-10
  connecting 11-13
  FLORA board 8, 9
  testing 11-13

W

watch
  building 163-167
  component, requisites 162
  creating, summarizing steps 178-181
  design 167-169
  leather, stitching 173-177
  notifications, receiving 181-187
  pieces, connecting 172, 173
  soldering 167-169
wearables 2, 3

Thank you for buying
Arduino Wearable Projects

About Packt Publishing

Packt, pronounced 'packed', published its first book, *Mastering phpMyAdmin for Effective MySQL Management*, in April 2004, and subsequently continued to specialize in publishing highly focused books on specific technologies and solutions.

Our books and publications share the experiences of your fellow IT professionals in adapting and customizing today's systems, applications, and frameworks. Our solution-based books give you the knowledge and power to customize the software and technologies you're using to get the job done. Packt books are more specific and less general than the IT books you have seen in the past. Our unique business model allows us to bring you more focused information, giving you more of what you need to know, and less of what you don't.

Packt is a modern yet unique publishing company that focuses on producing quality, cutting-edge books for communities of developers, administrators, and newbies alike. For more information, please visit our website at www.packtpub.com.

About Packt Open Source

In 2010, Packt launched two new brands, Packt Open Source and Packt Enterprise, in order to continue its focus on specialization. This book is part of the Packt Open Source brand, home to books published on software built around open source licenses, and offering information to anybody from advanced developers to budding web designers. The Open Source brand also runs Packt's Open Source Royalty Scheme, by which Packt gives a royalty to each open source project about whose software a book is sold.

Writing for Packt

We welcome all inquiries from people who are interested in authoring. Book proposals should be sent to author@packtpub.com. If your book idea is still at an early stage and you would like to discuss it first before writing a formal book proposal, then please contact us; one of our commissioning editors will get in touch with you.

We're not just looking for published authors; if you have strong technical skills but no writing experience, our experienced editors can help you develop a writing career, or simply get some additional reward for your expertise.

Arduino Essentials

ISBN: 978-1-78439-856-9 Paperback: 206 pages

Enter the world of Arduino and its peripherals and start creating interesting projects

1. Meet Arduino and its main components and understand how they work to use them in your real-world projects.

2. Assemble circuits using the most common electronic devices such as LEDs, switches, optocouplers, motors, and photocells and connect them to Arduino.

3. A Precise step-by-step guide to apply basic Arduino programming techniques in the C language.

Python Programming for Arduino

ISBN: 978-1-78328-593-8 Paperback: 400 pages

Develop practical internet of Things prototypes and applications with Arduino and Python

1. Transform your hardware ideas into real-world applications using Arduino and Python.

2. Design and develop hardware prototypes, interactive user interfaces, and cloud-connected applications for your projects.

3. Explore and expand examples to enrich your connected device's applications with this step-by-step guide.

Please check **www.PacktPub.com** for information on our titles

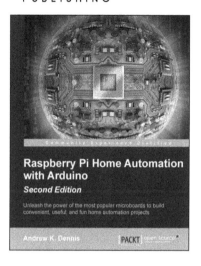

Raspberry Pi Home Automation with Arduino
Second Edition

ISBN: 978-1-78439-920-7   Paperback: 148 pages

Unleash the power of the most popular microboards to build convenient, useful, and fun home automation projects

1. Revolutionize the way you automate your home by combining the power of the Raspberry Pi and Arduino.

2. Build simple yet awesome home automated projects using an Arduino and the Raspberry Pi.

3. Learn how to dynamically adjust your living environment with detailed step-by-step examples.

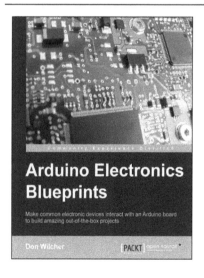

Arduino Electronics Blueprints

ISBN: 978-1-78439-360-1   Paperback: 252 pages

Make common electronic devices interact with an Arduino board to build amazing out-of-the-box projects

1. Build interactive electronic devices using the Arduino.

2. Learn about web page, touch sensor, Bluetooth, and infrared controls.

3. A project-based guide to create smartly interactive electronic devices with the Arduino.

Please check **www.PacktPub.com** for information on our titles

www.ingramcontent.com/pod-product-compliance
Lightning Source LLC
Chambersburg PA
CBHW060555060326
40690CB00017B/3718